SECURITY MANAGEMENT FOR OCCUPATIONAL SAFETY

Occupational Safety and Health Guide Series

Series Editor

Thomas D. Schneid
Eastern Kentucky University
Richmond, Kentucky

Published Titles

The Comprehensive Handbook of School Safety, *E. Scott Dunlap*

Corporate Safety Compliance: OSHA, Ethics, and the Law, *Thomas D. Schneid*

Creative Safety Solutions, *Thomas D. Schneid*

Disaster Management and Preparedness, *Thomas D. Schneid and Larry R. Collins*

Discrimination Law Issues for the Safety Professional, *Thomas D. Schneid*

Labor and Employment Issues for the Safety Professional, *Thomas D. Schneid*

Loss Control Auditing: A Guide for Conducting Fire, Safety, and Security Audits, *E. Scott Dunlap*

Loss Prevention and Safety Control: Terms and Definitions, *Dennis P. Nolan*

Managing Workers' Compensation: A Guide to Injury Reduction and Effective Claim Management, *Keith R. Wertz and James J. Bryant*

Motor Carrier Safety: A Guide to Regulatory Compliance, *E. Scott Dunlap*

Occupational Health Guide to Violence in the Workplace, *Thomas D. Schneid*

Physical Hazards of the Workplace, *Larry R. Collins and Thomas D. Schneid*

Safety Performance in a Lean Environment: A Guide to Building Safety into a Process, *Paul F. English*

Security Management: A Critical Thinking Approach, *Michael Land, Truett Ricks, and Bobby Ricks*

Forthcoming Titles

Workplace Safety and Health: Assessing Current Practices and Promoting Change in the Profession, *Thomas D. Schneid*

SECURITY MANAGEMENT FOR OCCUPATIONAL SAFETY

MICHAEL LAND

CRC Press
Taylor & Francis Group
Boca Raton London New York

CRC Press is an imprint of the
Taylor & Francis Group, an **informa** business

CRC Press
Taylor & Francis Group
6000 Broken Sound Parkway NW, Suite 300
Boca Raton, FL 33487-2742

First issued in paperback 2019

© 2014 by Taylor & Francis Group, LLC
CRC Press is an imprint of Taylor & Francis Group, an Informa business

No claim to original U.S. Government works

ISBN-13: 978-1-4665-6120-5 (hbk)
ISBN-13: 978-0-367-37914-8 (pbk)

Library of Congress Cataloging-in-Publication Data

Land, Michael (Industrial engineer)
   Occupational safety management : a critical thinking approach / Michael Land.
      pages cm. -- (Occupational safety & health guide series ; 15)
   Includes bibliographical references and index.
   ISBN 978-1-4665-6120-5 (hardback)
   1. Industrial safety--Management. I. Title.

T55.L285 2013
363.11'06--dc23   2013026448

Visit the Taylor & Francis Web site at
http://www.taylorandfrancis.com

and the CRC Press Web site at
http://www.crcpress.com

Contents

Preface

Safety is a paradox. Although we all want it, it is often viewed as intrusive, a hassle, or something that limits our personal, if not professional, freedoms. At the same time, if we need safety, security, and protection, we can never have enough. It is the intention of this book to provide an occupational safety practitioner with the ability to critically examine his or her organizational environment and provide a mechanism to make it safer while assuming the best possible relationship between obtrusion and necessity.

Occupational management is a very diverse function for which safety practitioners must plan, manage people, administer budgets, process information, as well as prepare for emergencies, violence, and other loss scenarios. This book is written for students who would like to become occupational safety managers in their professional pursuits, as well as for the current manager who wants to take a more critical and comprehensive approach to safety. *Occupational Safety Management: A Critical Thinking Approach* provides a methodical process to look at safety management functions and challenges.

This book presents the most accurately balanced picture of safety functions by using a critical thinking approach to interpret data as a tool for providing more effective occupational safety management. It is not the intent of this book to attempt to instruct someone on how to secure, guard, or protect with technical proficiency. Although there are commonalities in many aspects of occupational risks and hazards, all are going to be unique. The theme of this book is to create a practitioner who will completely examine the environment and make informed, well-thought-out judgments to tailor a security program to fit a specific organization.

This book enables students to think clearly and critically about the process of security management. It emphasizes the ability to articulate the differing aspects of business and security management by reasoning through complex problems in the changing organizational landscape. This book guides students through the stages of the critical thinking process to formulate a practical program for security management.

The book emphasizes the core security management competencies of planning, organizing, staffing, and leading while providing a process for critically analyzing those functions. The book stresses the benefits of using a methodical critical thinking process in building a comprehensive safety management system. The book specifically addresses information security, cybersecurity, energy sector security, chemical security, and general security management while utilizing a critical thinking framework.

This book goes further than other books that are available regarding security management in that it not only provides fundamental concepts in security but also will help create informed, critical, and creative security managers who communicate effectively in their environment.

About the Author

Michael Land, PhD, is a student success specialist for Eastern Kentucky University, College of Justice & Safety. In addition to his more than 20 years at EKU, he has consulted for both state and federal agencies in different facets of safe and secure technology design and implementation. Dr. Land has taught many courses in security, occupational safety, and loss prevention throughout his career at EKU. Dr. Land holds a BS in security and loss prevention and an MS in loss prevention administration, both from Eastern Kentucky University, and a doctorate in educational leadership from Lincoln Memorial University.

1 Introduction to Occupational Safety Management

Occupational safety is about creating a safe work environment, as well as an understanding, to facilitate the positive physical, mental, and social well-being of workers. A common definition of *occupational health* was adopted by the Joint International Labor Organization and World Health Organization Committee on Occupational Health at its first session in 1950, and it was revised at its twelfth session in 1995. The original is as follows:

> Occupational health should aim at: the promotion and maintenance of the highest degree of physical, mental and social well-being of workers in all occupations; the prevention amongst workers of departures from health caused by their working conditions; the protection of workers in their employment from risks resulting from factors adverse to health; the placing and maintenance of the worker in an occupational environment adapted to his physiological and psychological capabilities; and, to summarize, the adaptation of work to man and of each man to his job. (Joint ILO/WHO Committee on Occupational Health, 1957)

In an organization, occupational safety is important for moral, legal, and financial reasons. All organizations have a duty of care to ensure that employees and others who may be affected by the company's undertakings remain safe at all times. Moral obligations involve the protection of employees' lives and health. Legal reasons for occupational safety and health (OSH) practices relate to the preventative, punitive, and compensatory effects of laws that protect workers' safety and health. OSH can also reduce employee injury- and illness-related costs, including medical care, sick leave, and disability benefit costs. Furthermore, the concept of a safe working culture is to create an understanding of the safety functions, which facilitates their diffusion and adoption by organizational staff. The safety culture is reflected in practice in organizational systems, policies, principles, training, and quality management.

Based on these common understandings, OSH has three primary objectives:

1. The maintenance and promotion of the health and working ability of staff
2. The improvement of the working environment and work to become more conducive to safety and health
3. The development of work organizations and work cultures in a direction that supports health and safety at work, and in doing so also promotes a positive social climate and smooth organizational operation

In the United States, the Occupational Safety and Health Act of 1970 created both the National Institute for Occupational Safety and Health (NIOSH) and the Occupational Safety and Health Administration (OSHA). OSHA, which is part of the U.S. Department of Labor, is responsible for developing and enforcing workplace safety and health regulations. NIOSH, in the U.S. Department of Health and Human Services, is focused on research, information, education, and training in OSH.

The main tasks undertaken by the OSH manager include

- Develop processes, procedures, criteria, requirements, and methods to attain the best possible management of the hazards and exposures that can cause injury to people and/or damage property or the environment.
- Apply good business practices and economic principles for the efficient use of resources to add to the importance of the safety processes.
- Encourage other members of the company to contribute by exchanging ideas and different approaches to make sure that everyone in the corporation possesses OHS knowledge and has functional roles in the development and execution of safety procedures.
- Assess services, outcomes, methods, equipment, workstations, and procedures by using qualitative and quantitative methods to recognize the hazards and measure the related risks.
- Examine all possibilities, effectiveness, reliability, and expenditure to attain the best results for the company concerned.

The knowledge required for an OSH manager in the United States includes

- Constitutional and case law controlling safety, health, and the environment
- Operational procedures to plan and develop safe work practices
- Safety, health, and environmental sciences
- Design of hazard control systems (i.e., fall protection and scaffoldings)
- Design of recordkeeping systems that take collection into account, as well as storage, interpretation, and dissemination
- Processes and systems for attaining safety through design

Other skills required by an OSH manager in the United States include

- Understanding and relating to systems, policies, and rules
- Holding checks and having control methods for possible hazardous exposures
- Mathematical and statistical analysis
- Examining manufacturing hazards
- Planning safe work practices for systems, facilities, and equipment
- Understanding and using safety, health, and environmental science information for the improvement of procedures
- Interpersonal communication skills

Organizational employees want a safe workplace. For organizational leadership, ignoring dangers and risks is unacceptable. Being proactive directs us to look at our individual communities, identify its assets, recognize potential targets, and develop a strategy to prioritize and plan against events, whether human-made or natural.

Through risk assessment and management, private enterprise identifies hazards and weighs the risk of not providing an acceptable level of safety against the results should a loss event occur. For example, the decision to implement a policy to indemnify a hazard is weighed against the probability of the event occurring and the consequences of the event. In addition to threats, employee safety affects morale, which in turn impacts productivity.

Private enterprise spends billions of dollars on occupational safety in order to provide a safe workplace. Governmental laws or regulations, insurance requirements, and partnership agreements may cause an enterprise to maintain a particular level of safety, which in turn fuels the need for occupational safety mechanisms, personnel, and education. Occupational safety is about risk reduction. Whether it is for a government agency or a private enterprise, occupational safety focuses on three main ideas: the maintenance and promotion of the health and working ability of staff, improvement of the working environment in order to become more conducive to staff safety and health, and the development of work organizations and working cultures in a direction that supports health and safety at work and in doing so also promotes a positive social climate and smooth organizational operation.

Assuring a workplace where employees feel safe affords them a comfort level that allows them to work at peak efficiency without fear. Providing a safe and secure facility leads to increased productivity, which lowers costs and raises profits. For example, an apartment or office complex provides for a safe environment to attract tenants, and a retail store provides a safe environment to attract customers. If tenants or customers do not feel safe, they will take their business elsewhere. In addition to these benefits, the risk of loss through lawsuits is lessened. A facility that implements and maintains a level of occupational safety lowers the risk of lawsuit and the amount of a judgment in the event of a lawsuit.

The occupational safety manager must be good at managing and leading. The occupational safety manager must possess a measure of technical knowledge, be able to work with other people, and be able to develop and conceptualize ideas. The degree of each skill that is required varies with the position and role of the occupational safety manager. First-line supervisors work closely with the line staff, need more technical skill regarding the work activity, and would use little conceptual skill. Conversely, executives would spend more time using their conceptual skills, rather than their knowledge of technical operations, to move the organization forward. A midlevel manager would possess some technical and some conceptual skills in roughly equal amounts. Again, the ability to work with others is equally necessary at all levels of management.

A good occupational safety manager also needs leadership skills. We manage things; we lead people. Managing operations involves focusing on safe processes and process control, while leaders focus on the people and creating the organizational culture. Leadership is defined as "a process whereby an individual influences a group of individuals to achieve a common goal" (Northouse, 2006). What makes a good and effective leader is the ability to influence others. The leader sets the vision of the organization or group. The leader "sells" the vision to the group to inspire and motivate the group to achieve specific goals. The leader communicates the goals to the group and makes clear the expectations in order to achieve the goals. Leaders seek commitment from the group to strive for these goals. Leaders empower the group with the authority necessary to complete the tasks.

Managers coordinate assignments and provide the resources to allow their staff to perform their assignments efficiently and effectively. They provide structure and outline processes necessary to maximize productivity and to prevent waste. Managers establish policies and rules to provide structure for defining work procedures.

Managers are problem solvers. They examine work processes and identify ways to eliminate unnecessary tasks and streamline the workflow.

HOW TO USE THIS BOOK

Management and leadership skills can be learned. This book will explore fundamental concepts of both leadership and management, and then provide application of these concepts to the role of the occupational safety manager. This chapter is like a macropedia; each chapter listed here will lead you to key information.

Chapter 2 explores the fundamentals of critical thinking and analysis. Managers must be ready to analyze problems and understand how information is influenced by bias and perceptions, even manipulation. Chapter 3 examines the core competencies of the occupational safety manager. Chapter 4 tells us how to build an occupational safety management program. The remaining chapters are as follows:

Chapter 5: Developing Policies and Procedures for Occupational Safety Management
Chapter 6: Staffing and Occupational Safety Management
Chapter 7: Enterprise Risk Management and Occupational Safety
Chapter 8: Occupational Safety Leadership
Chapter 9: Comprehensive Risk Assessment for the Occupational Safety Manager
Chapter 10: Computer and Information Security for Occupational Safety
Chapter 11: Cybersecurity for Occupational Safety
Chapter 12: Occupational Safety Investigations
Chapter 13: Safety and Security Management for Chemical Facilities
Chapter 14: Safety and Security Management for the Energy Sector
Chapter 15: Contemporary Safety Management

Law and liability, threat and risk assessment, emergency preparedness, and safety design form the beginning of identifying a need for occupational safety awareness. The basic questions of "Where we are?" "Where do we need to be?" and "How do we get there?" identify occupational safety needs and help us to form our objective for efficient and effective occupational safety programs. Risk assessment and cost analysis round out the formula for determining occupational safety needs.

This book examines the policies and procedures necessary for an efficient and effective occupational safety program. For all occupational safety systems to work, the overall occupational safety policy must address issues of access control; employee, visitor, and package screening; and deliveries. Awareness programs must be developed to educate personnel on threats and occupational safety procedures. Finally, employee background checks and visitor access requirements are examined to round out the occupational safety program.

As Dale Carnegie famously said, "If you do the little jobs well, the big ones will tend to take care of themselves." Too many times, we look at the "big picture" but fail to see the little things. We are prepared for the terrorist attacks, but ignore incidents we see as minor distractions or nuisances. The broken windows theory teaches us that small items like broken windows, graffiti, unkempt lawns, and the like invite small crimes like trespassing and criminal mischief. Such activity, if unheeded, evolves into larger activity until we slide down a slippery slope and find ourselves in a quagmire. Attention to detail in handling minor incidents will prevent many major incidents while preparing us for the ones we cannot stop.

EXERCISE

You are assuming a new position of occupational safety manager in a relatively new company that has never had an occupational safety department before. Although some members of management understand the need for a formal occupational safety department, others do not. The belief of the latter group tends to be "Sure, there have been safety issues in the company, but the role of occupational safety is not going to be without cost, either." Is it your first job to meet with the departmental managers and explain to them why their organization needs occupational safety? Is occupational safety management an investment or just a requirement?

REFERENCES

Joint ILO/WHO Committee on Occupational Health. (1957). World Health Organization Technical Report Series No. 135. Accessed May 21, 2013, at http://www.who.int/occupational_health/publications/ILO_WHO_1957_report_of_the_joint_committee.pdf
Northouse, P. (2006). *Leadership: Theory and Practice.* Thousand Oaks, CA: Sage Publications.

2 Occupational Safety Management and Critical Thinking

Critical thinking is a logical process of examining data to gain an understanding to guide personal belief and action. Critical thinking provides modern occupational safety managers with a powerful tool to utilize when fulfilling the requirements of their position. However, literature on occupational safety management and critical thinking methodologies is sparse. Very few researchers and practitioners have identified the value of critical thinking in addressing the issues of occupational safety.

Critical thinking is a tool to move the safety manager past personal bias, or the biases of others, to provide objectivity in the work environment. Safety practitioners know that safety management is rooted in methodical processes that serve as a guide for providing a safe working environment. As such, many safety processes already utilize a homogenized type of critical thinking. Critical thinking for the safety manager also facilitates reflective thinking, in which the concept of a safe working culture is to create an understanding of safety functions, thus facilitating their diffusion and adoption by organizational staff. Such a culture is reflected in practice in organizational systems, policies, principles, training, and quality management.

The critical nature of occupational safety management tools such as risk assessments and criticality and vulnerability studies reflects just a glimpse of the methodical tools of the occupational safety manager. The risk assessment itself looks at the value of organizational assets, accounting for both the criticality and vulnerability of the asset to determine the amount of the occupational safety investment. In itself, the comprehensive process of risk analysis utilizes a process of critical thinking.

The intent of this chapter is to move beyond statistical safety tools and focus on ways that the occupational safety manager can apply critical thinking skills to improve an organization's everyday performance. This chapter provides an understanding of the critical thinking process as a methodical process that the safety manager should cultivate as an interpretive lens through which to evaluate the occupational safety environment on a daily basis.

For the occupational safety manager, there is nothing more practical than a methodical process of thinking to reduce risk or loss. For managers, good thinking pays off while poor thinking causes problems, wasting resources and time. Critical thinking allows occupational safety managers to envision their duties in a logical process while focusing on making decisions and solving problems.

Critical thinking should become a natural process to occupational safety managers. Schon's (1983) idea of critical thinking involves understanding the relations between theoretical and practical knowledge as part of our daily spontaneous and intuitive management processes. These spontaneous-intuitive processes become a reflection process with which the occupational safety manager can instinctively, yet critically, interpret a situation.

Essential for thinking about how occupational safety managers use critical reflection is reflected in Schon's (1983) description of them as researchers and practitioners who can theorize and know their actions because of reflection. Occupational safety managers experience revelation, perplexity, or uncertainty in situations that they find unknown or unique. Occupational safety managers reflect on a given situation and on prior understandings that have been implicit in their behavior. Occupational safety managers instinctively carry out an intuitive methodical process that serves to generate both a new understanding of the phenomena and a change in the situation.

In the reflective process, occupational safety managers do not separate thinking from doing, paving the way to a decision that they must later convert to action. The occupational safety manager is a practitioner, a researcher who can engage his or her safety situations by bringing thinking and action together, by interactively and critically engaging the relations of means and ends in situations that are unique, difficult, and uncertain.

Critical thinking and reflection are much more than a cognitive process of analysis. Critical thinking is more than a form of technical rationality to a potential loss situation. Instead, critical thinking and reflection become an artistic process (Brookfield, 1987, p. 155). During their work performance, occupational safety managers may be faced with unexpected and unfamiliar situations that they need to respond to. They need to look at a situation and engage critical thinking so that the situation becomes a part of the collection of experiences they use to build theories and responses. Occupational safety managers need to observe, research, and reflect on processes that impact the workplace. They also need to be able to articulate their reflection-in-action; otherwise, they cannot instill the process in others (Schon, 1983, p. 243). All thinking and practices demand a stop-and-think period, that is, time for reflection. The occupational safety manager should hone a methodology, bringing critical thinking, reflection, and action dynamically together, leading to superior daily performance.

BACKGROUND OF CRITICAL THINKING

To arrive at a process of critical thinking that could best serve an occupational safety manager, examination must be provided for a better understanding of critical thinking processes. Critical thinking has been expressed and defined in several ways. In the literature, there are several definitions and ways of conceptualizing critical thinking. Chance (1986) defines critical thinking as the ability to analyze facts, generate and organize ideas, defend opinions, make comparisons, draw inferences, evaluate arguments, and solve problems. Tama (1989) defines it as a way of reasoning that demands adequate support for one's beliefs and an unwillingness to be persuaded unless support is forthcoming.

In Mayer and Goodchild (1990), critical thinking is an active, systematic process of understanding and evaluating arguments. An argument provides an assertion about the properties of some object or the relationship between two or more objects and evidence to support or refute the assertion. Critical thinkers recognize that there is no one correct way to understand and evaluate influences and that not all attempts are successful.

Although critical thinking receives much attention in the higher education environment, the idea is not new. John Dewey, the proponent of the progressive education movement, is often regarded as the father of modern critical thinking. In *How We Think* (1910), Dewey calls critical thinking reflective thought and defines it as "active, persistent, and careful consideration of a belief or supposed form of knowledge in light of the grounds which support it and the further conclusions for which is tends" (9). Dewey recognizes that what matters are the reasons we have for believing something and the implications of our beliefs (Fisher, 2001).

For Dewey, critical thinking is essentially an active process in which you think things through yourself, raise questions yourself, find relevant information yourself, and solve problems yourself, rather than learning in a largely passive way from someone else. Thus, to develop critical thinking skills, individuals must be active learners in the learning process, and they must identify and solve unstructured problems using multiple information sources.

An occupational safety manager must be able to deal with the unstructured situations that arise in the work environment. If every problem or issue that a manager faced could be dealt with by using the same structured response, managers would not need to "think" but to merely apply their rubric to the situation and then go to the next issue. However, that scenario seldom exists in the occupational safety environment. While many facets of occupational safety can be addressed through proactive management tools, there is still a huge variation of problems that the occupational safety manager deals with on a daily basis.

Dewey roots critical thinking in engagement with a problem (Dewey, 1916). A century later, contemporary critical thinking researchers promote using context, elements, and disciplines to gain insight, and understanding issues in a deep way, to help one make better decisions in life (Niosich, 2012). For Niosich, critical thinking allows individuals to more effectively interpret information, keep important information, and actually retain that information for future use and reference with more ease over time since critical thinking skills allow a person to become an active listener rather than a passive recipient of information.

Niosich (2012) uses an academic approach describing the technical process of critical thinking to assist improving individual capabilities through critical thinking. Niosich's point of view is that with practicing critical thinking, we will have the following outcomes:

Whether it is in writing or reading, in analysis or evaluation, in the discipline as well as in your life outside school, critical thinking creates value. It takes effort, especially before you get used to it. But it has clear practical benefits that far exceed the effort. It will produce better answers, better grades, in more courses, in more professions, with ultimately less work, than any alternative. More than that, it gives insight that can make your life richer, by bringing the elements, the standards, and the disciplines into learning to think things through. (p. 195)

If all parts of critical thinking are understood and utilized, then the occupational safety manager should be checking for accuracy, identifying assumptions, drawing relevant conclusions, and thinking questions out in terms of the fundamental and powerful concepts of the discipline. Critical thinking is reflective. It is different from just thinking. It is metacognitive, which means that it involves thinking about your thinking.

Critical thinking involves standards used as a measure for thinking. Some examples of standards are accuracy, relevance, and depth. Niosich states that "there are no rules that guarantee our thinking will be correct," so the critical thinker must allow for "self-correcting" by evaluating the reasoning (Niosich, 2012, p. 5). The three parts of critical thinking emphasized are "asking questions … that go to the heart of the matter," "trying to answer those questions by reasoning them out," and "believing in the results of our reasoning" (p. 5).

The elements of reasoning (keeping in mind the context and alternatives) are, in no particular order as this is a circle, the point of view, the purpose, the question at issue, assumptions, implications and consequences, information, concepts, conclusions, and interpretations (Niosich, 2012, pp. 49–60).

- Purpose is having an objective.
- Question at issue can be thought of as the problem being addressed.
- Assumptions are everything you take for granted when you think through something.
- Implications and consequences fall into the area beyond the end of critical thinking.
- Information is the data, evidence, and observations used during reasoning. Reliable information from numerous sources should be used.
- Concepts are our understanding of a term or issue.
- Conclusions are the decisions at which we arrive.
- Point of view is the perspective used to address an issue.
- Alternatives are simply other possibilities.
- Context is similar to the setting or background of reasoning.

There are seven standards to thinking critically: clearness, accuracy, importance or relevance, sufficiency, depth, breadth, and precision (Niosich, 2012, pp. 133–148).

- Clear thinking is when you can state your meaning exactly, when you can elaborate on it and explain it, and when you can give good examples and illustrations of it.
- Importance occurs when the information is directly relevant to addressing the problem at hand.
- Sufficiency happens when you've thought about a question until you've reasoned it out thoroughly enough for the purpose at hand, when it is adequate for what is needed, and when you've taken account of all necessary factors.
- The goal for depth and breadth is to develop an intuitive feel for when it is important to delve deeper into an issue and when it is important to look at it more broadly by taking account of other related issues.
- What is precise will always be relative both to the purpose of the reasoning and to the context.

Niosich describes the core process as addressing a question or problem, thinking it through using the elements of reasoning, reasoning out all aspects of the issue through the lens of the discipline when appropriate, and monitoring reasoning using critical thinking standards (2012, p. 169). Evaluation, comparing and contrasting, application, decision making, action, and living mindfully help the critical-thinking process (p. 172).

"Critical thinking transfers" (Niosich, 2012, p. xxviii). With "time, practice, and commitment," the reader will be able to use the material supporting critical thinking and apply critical thinking to improve not just in school but in all decision making, with improved decisions as a result. However, we must take action, or this will not be fruitful.

Whether it is in writing or reading, in analysis or evaluation, or in the discipline as well as in your life outside school, critical thinking creates value. It takes effort, especially before you get used to it. But it has clear practical benefits that far exceed the effort. It will produce better answers and better outcomes in more courses and more professions, and with ultimately less work than any alternative. More than that, it gives insight that can make your life richer by bringing the elements, the standards, and the disciplines into learning to think things through (Niosich, 2012, p. 195). So, the material will allow the reader to incorporate critical thinking into daily life experiences with improved decision-making results.

Paul and Elder (2005) argue that critical thinking involves the ability to raise vital questions and problems; to gather and assess relevant information; to use abstract ideas to interpret information effectively; to come to well-reasoned conclusions and solutions, testing them against relevant criteria or standards; and to think open mindedly within alternative systems of thought, recognizing and assessing their assumptions, implications, and practical consequences.

Paul and Elder (2005) further argue that successful thinkers move more or less sequentially through a standard process of identifying problems, making reasonable assumptions about the nature of the problems, discerning criteria according to which information about the problems can be deemed relevant and be well understood, making inferences from the pertinent data, and organizing these inferences into concepts that will help them come up with a workable solution.

Paul and Elder (2005) use these "Elements of Thought" to create a checklist for students that will guide them in their analytic thinking; it is shown in Table 2.1. The benefit of this model and checklist to instructors and students is that they teach individuals how to analyze a broad range of materials, from news articles to chapters in textbooks, government reports, novels, and poems.

Students should regularly use the following checklist for reasoning to improve their thinking in any discipline or subject area:

- All reasoning has a purpose.
  - State your purpose clearly.
  - Distinguish your purpose from related purposes.
  - Check periodically to be sure you are still on target.
  - Choose significant and realistic purposes.

- All reasoning is an attempt to settle some question, figure something out, or solve some problem.
  - State the question at issue clearly and precisely.
  - Express the question in several ways to clarify its meaning and scope.
  - Break the question into subquestions.
  - Distinguish questions that have definitive answers from those that are a matter of opinion and from those that require consideration of multiple viewpoints.
- All reasoning is based on data, information, and evidence.
  - Restrict your claims to those supported by the data you have.
  - Search for information that opposes your position and information that supports it.
  - Make sure that all information used is clear, accurate, and relevant to the question at issue.
  - Make sure you have gathered sufficient information.
- All reasoning contains inferences or interpretations by which we draw conclusions and give meaning to data.
  - Infer only what the evidence implies.
  - Check inferences for their consistency with each other.
  - Identify assumptions that lead you to your inferences.
- All reasoning is expressed through, and shaped by, concepts and ideas.
  - Identify key concepts and explain them clearly.
  - Consider alternative concepts or alternative definitions of concepts.
  - Make sure you are using concepts with care and precision.
- All reasoning is based on assumptions (beliefs you take for granted).
  - Clearly identify your assumptions and determine whether they are justifiable.
  - Consider how your assumptions are shaping your point of view.
- All reasoning is done from some point of view.
  - Identify your point of view.
  - Seek other points of view and identify their strengths and weaknesses.
  - Strive to be fair-minded in evaluating all points of view.
- All reasoning leads somewhere or has implications and consequences.
  - Trace the implications and consequences that follow from your reasoning.
  - Search for negative as well as positive implications.

The point of critical thinking and occupational safety management is to promote the active exploration of ideas. Critical thinking should be developed in occupational safety managers as they move through the process of identifying problems, gathering facts and data about problems, making reasonable assumptions about the nature of the problems, discerning criteria to analyze the problems, and identifying possible solutions to complex problems and their consequences.

Occupational safety managers use tools such as risk assessments, criticality and vulnerability studies, return on program investment analyses, and cost–benefit analyses to gain an understanding of loss environments. In itself, the comprehensive process of risk analysis utilizes a process of critical thinking. However, the intent of this chapter has been to embrace the formal processes of critical thinking and to make critical thinking a reflective part of the everyday role of occupational safety management.

CHAPTER QUESTIONS

1. Summarize five major points made in this chapter.
2. Discuss the essence of this chapter using a metaphor.
3. Explain critical thinking to your neighbor, who has a high school education and has not been in the work force for fifteen years. What assumptions did you make when you were developing your explanation, and why?
4. How might the information you gained from this reading on critically thinking affect you personally and professionally?

REFERENCES

Brookfield, S. D. (1987). *Developing Critical Thinkers: Challenging Adults to Explore Alternative Ways of Thinking and Acting*. New York: Jossey-Bass.

Chance, P. (1986). *Thinking in the Classroom: A Review of Programs*. New York: Instructors College Press.

Dewey, J. (1916). *Democracy and Education*. New York: Macmillan.

Dewey, J. (1920). *How We Think*. Boston: D.C. Heath.

Erskine, J. A., Michiel, R. L., & Mauffette-Leenders, L. A. (1981). *Teaching with Cases*. Waterloo, Canada: Davis and Hedersen.

Fisher, A. (2001) *Critical Thinking: An Introduction*. Cambridge, UK: Cambridge University Press.

Mayer, R., & Goodchild, F. (1990). *The Critical Thinker*. New York: Wm. C. Brown.

Niosich, G. (2012). *Learning to Think Things Through: A Guide to Critical Thinking across the Curriculum*. 4th ed. Upper Saddle River, NJ: Prentice Hall.

Paul, R., & Elder, L. (2005). *The Miniature Guide to Critical Thinking: Concepts and Tools*. Tomales, CA: Foundation for Critical Thinking.

Schon, D. (1983). *The Reflective Practitioner: How Professionals Think in Action*. New York: Basic Books.

Tama, C. (1989). Critical thinking has a place in every classroom. *Journal of Reading*, 33, 64–65.

3 Core Competencies of the Occupational Safety Manager

The core competencies of the occupational safety (OS) manager include the ability to develop processes, procedures, and methods to attain the best possible management of the risks that can cause injury to people, and damage property or the environment. The OS manager should apply good business practices and economic principles for the efficient use of resources to add to the importance of the safety processes.

The OS manager should encourage other members of the organization to contribute by exchanging ideas and other different approaches to make sure that everyone in the corporation possesses OS knowledge and has functional roles in the development and execution of safety procedures. The OS manager must assess services, outcomes, methods, equipment, workstations, and procedures by using qualitative and quantitative methods to recognize the hazards and measure the related risks. In addition, he or she must examine all possibilities, effectiveness, reliability, and expenditures to attain the best results for the company concerned.

PLANNING, ORGANIZATION, MOTIVATION, AND CRITICAL ANALYSIS

PLANNING

Every organization, whether small or large, or public or private, must have a road map or guide as to where they want to go or believe that they are going. This must start by knowing what has happened in the past, what is occurring in the present, and where they want to go in the future. This necessitates having a plan. Likewise, this is true for OS management.

What is a plan? A plan is selecting and relating facts, and making and using assumptions, regarding the future in a rational course of action. The planning process is related to rational decision making when a course or courses of action must be made. Assumptions generally are statements or beliefs that are accepted as true without proof or demonstration. These assumptions can be considered a bundle of decisions about what is known about the past and will be included in a plan when preparing for the future by making decisions now. A plan represents expenditures of thought, time, and knowledge now for an investment in the future.

The rational selection of a course of action (i.e., the making of a rational plan) includes basically the same procedures as those of any rational decision. This means that most, and if possible all, of the courses of action must be identified, the consequences of each course (area) must be predicted or known in advance of any decision making, and the courses (onward movement in a particular direction) having the preferred results or consequences must or should be selected. In order to achieve the desired results, the planning process should comprise the following five activities: problem identification or analysis of the situation, goal setting (for a desired future state of affairs), design of courses of action (alternative approaches to goal attainment), comparative evaluation of consequences (predicting the results of each alternative), and final selection of a course of action (decision making).

These five activities create a dynamic process that involves a number of courses of action or methods for generating plans that provide an organization with guides to sustain renewal and change in terms of more effective goal accomplishment. Combining these five activities denotes that progressive actions are being performed by persons in the course of moving the organization from one state or situation to another. Combining these activities creates a process that allows or requires a flow of interrelated events moving toward some goal, purpose, or end. *Flow* implies a movement through time in the direction of a consequence. *Interrelated* denotes interactions within the process and events that are highly relevant one to another. *Events* are changes or happenings that occur at one point of time, and they may be any of an infinite number of phenomena. *Goals* suggest a decision maker's objective, while *purpose* suggests either human objectives or objectives in a nonmaterial sense. *End* implies some conclusion or consequence, which may not necessarily be sought or planned by the decision maker. This explanation or definition of process may or may not have human-intended consequences.

It is certainly true that some goals can be achieved with relatively little planning. However, today, where many tasks have become quite complex, where more technology is involved, where more people want and demand to be informed and to participate in what's going to be done, and with an ever increasing diversity of personnel, products, and services, planning has become a necessity.

One may question the relevance of planning or a plan drawn from and is a major concept from classical management writings. One would have justification in trying to defend not having a plan at all. As for other specific points that a plan should have, which turn on the nature, importance, and condition of the organization for which the plan is drawn up, there could be a possibility of settling them beforehand by utilizing acceptable components of a known successful organizational plan.

ORGANIZATION

Organization can be defined as the establishing of a system of effective and structural behavioral or interpersonal relations among individuals, with the individuals being differentiated in terms of authority, status, and role, and with the purpose of achieving some goal or objective. Organizing results in an organization structure that provides a framework whereby human beings can favorably unite their efforts.

Classical organization theory is built around four primary concepts that were formalized from past successful occurrences. They are the division of labor, scalar and functional processes, structure, and the span of control.

1. The division of labor has been accepted as the cornerstone of the four concepts. From it, the other concepts flow as a natural consequence or effect. Growth within the organization requires the scalar and functional processes to necessitate specialization and departments of function. Organization structure is always dependent upon the direction of specialization development. Finally, a span of control problems results from the number of specialized functions under the authority and supervision of a manager.

2. Scalar and functional processes deal with the vertical and horizontal growth of the organization. The scalar process refers to the growth of the chain of command, the delegation of authority and responsibility, unity of command, and the obligation to report. The division of the organization into specialized parts and the regrouping of the parts into compatible units are matters pertaining to the functional process. This process focuses on the horizontal evolution of the line and staff in a formal organization. A strong belief has developed from the line–staff relationships in organizations that staff areas and positions were not to supervise and have authority over line processes.

3. Structure is the logical relationships of functions in an organization, arranged to accomplish the goals and objectives of the organization efficiently. Structure strongly implies system and pattern. Structure is the mechanism for introducing logical and consistent relationships among the diverse functions that comprise the organization.

4. The span of control concept relates to the number of subordinates a manager can effectively supervise. Are there numerical limitations to the subordinates one person can control? The answer is no. *Span* should refer to a number of persons, themselves carrying managerial and supervisory responsibilities, for whom the senior leader retains overarching responsibility in direction and planning, coordination, motivation, and control. Regardless of the interpretation of a span of control, a wide span yields a flat structure, whereas a short span results in a tall structure. In today's world, most leaders accept a wide span.

MOTIVATION AND CRITICAL ANALYSIS

From the four primary principles of classical management, especially span of control, comes the advent of centralization versus decentralization. Fredrick Winslow Taylor, who is accepted as the father of scientific management, had an engineering background; he had to deal with one location and one product, and how to organize an organization and its component parts in a more efficient and effect manner to make more profits. Emphasis was placed on arranging the organization in ways that would allow the workers to assist the organization. Under this system of management, the organization was primary and the workers were secondary.

In the United States during the 1870s through 1890s, the Industrial Revolution sprung up and products were not produced in one location or in one facility. Products such as steel, gas and oil, paint, chemicals, glass, and financial services were not available in only one location. Automobiles, railroads, and other products required numerous component parts. Organizations started exploring decentralization along with their centralized control.

For years, this type of span of control was best exemplified by the late Alfred P. Sloan at the General Motors Corporation (GM). Sloan's vision was to divide GM into as many parts as could be done consistently, place the most capable executive that could be found in charge of each part, and develop a system of coordination so that each part could strengthen and support the others. In academic circles, this was labeled "decentralized organization with centralized control." The units that exercised major control were finance, policy, and direct decisions that expanded all parts of the organization. Again, this led academia to foster a saying in any organization that if you don't control your budget and your personnel, then you don't control your part of the organization.

Again in the United States, from the 1870s to World War I, organizations grew to have large numbers of products produced at one or more locations by extremely large numbers of workers.

The emphasis by these organizations on developing the classical theories and in most cases disregarding the human aspect of the organization led to the neoclassical school of thought. This is a "tall" form of management.

Neoclassical theory embarked on the task of compensating for some of the deficiencies in classical doctrine. The neoclassical approach takes the postulates of the classical school, regarding the pillars of the organization, as givens. Neoclassical thought emphasized that these postulates were modified by people, acting independently or written in the context of the informed organization. This is a "flat" form of management.

Neoclassical thought had the support of behavioral scientists, and later mathematicians to assist in conducting studies and analyzing existing data. Studies of the repetitive nature of factory work, assembly-line (piece) work in time as workers completed and is a requirement and emphasis on how the informal organization effects these workers.

Without transparency and observable communication in an organization, an informal organization will develop. Five important aspects of informal organization as it operates within the formal organization are as follows:

1. Informal organizations act as agencies (parts) of social control. They generate a culture based on certain norms of conduct, which, in turn, demands conformity from group members. Diversity of personnel assists in reducing the effects that an informal culture can exert over an organization.
2. The form of human interrelationships in the informal organization requires techniques of analysis different from those used to plot the relationship in a formal organization.
3. Informal organizations have status and communication systems peculiar to themselves, and they are not necessarily derived from the formal organization.
4. The survival of informal organizations requires stable continuing relationships among the people in them. Informal organizations resist change. Change may involve learning a new job and developing relationships with new people.

5. The last aspect of analysis that appears to be central to the neoclassical view of the informal organizations is the study of the informal leader. A balancing act exists with the informal leader about how to assist and serve the informal group and at the same time assist and serve the formal organization. The informal leader and the formal leader must understand that the informal leadership position was not created by the formal organization.

In today's society, a 40-hour work week, with extended overtime, long vacation and sick time from work, 20 to 30 years of retirement for life at any age, and health and sick benefits fully covered are causing governmental organizations problems. These areas have and will create problems for future organizational leaders.

In regard to the motivation of most employees in any organization, two factors are important today and will continue to be important in the future:

1. Communication and transparency are important to all employees. The more they know about what is going on in their organization, the more they relate to the organization.
2. The involvement of employees at all levels and parts of the organization in organizational decision making will assist in the motivational aspects of employees. Authentic employee involvement must be the organizational norm. Employees expect that every suggestion they make for improving the organization will be considered. They want involvement; they want suggestions to be discussed with them and explanations given as to why or why not they will be implemented. Employees are not fools, and they do not want hypocrisy from management.

Learning theories as they relate to individual and group behaviors have evolved and developed since World War II. Drastic changes following World War II opened many opportunities for social science researchers to bring forth research work and studies, such as the Hawthorne study, from the years prior to World War II and to explore them as well as new ideas in a fast-paced society.

Two major researchers, Abraham H. Maslow and Fredrick Herzberg, developed models of motivation during this period. Maslow's hierarchy of needs (Maslow, 1954) is as follows, starting with the most basic needs at the bottom of the list and working up to the top:

Self-actualization and fulfillment
Esteem and status
Belonging and social needs
Safety and security
Physiological needs

This is a model of Herzberg's motivation–maintenance theory (Herzberg et al., 1959); again, this should be read from bottom to top:

Work itself, achievement, and possibility of growth and responsibility
Achievement, recognition, and status

Relations with supervisors, peer relations, relations with subordinates, and
   quality of supervision
Organization policy and administration, job security, and working conditions
Pay

These two models are cognitive theories of motivation that relate to the mental process by which knowledge is acquired. These models research how a worker obtains this knowledge through perception, reasoning, and intuition. A major difficulty with cognitive models of motivation is that they are not subject to precise scientific measurement and observation. One must argue that these two models are strong starting points if any type of reinforcement to modify behavior by its consequences is in place. One could also argue that most people today have satisfied their lower needs and are seeking to satisfy their higher level needs at work. Satisfying higher level needs can be done in most organizations by decreasing hierarchical control and overspecialization of roles. Many studies have discovered that decreasing hierarchical control leads to greater self-value and, hence, a greater degree of expectancy of success and value attainment.

Critical analysis is defining or stating a problem, and it must be the first order of business when analyzing a critical or potential problem before making a decision. The core or central part of the problem must be stated before an analysis starts and a decision is made. After stating the problem, decisions must be made as to what kind of data must be collected, and priorities need to be set on how a decision will be made.

In social science, we use number systems for essentially three purposes: (1) to classify things, (2) to order things, and (3) to quantify things. Since data are measures of such variables, we refer to them as (1) nominal data, (2) ordinal data, (3) interval data, and (4) ratio data.

Nominal data can be name, race, gender, or other data that does not have a numerical value. Interval data allows for varying values such as temperature, but not ratio values because they have scaled values. Ratio data is the value between data objects, such as height, weight, and age. Ordinal data refers to data with a ranking or ordering. Likert-scaled questions are an example of ordinal data.

Nominal data are defined as such whenever we assign numbers to a set of categories without reference to the direction or magnitude of difference among the alternatives. Nominal pertains to, or consists of, a name or names.

An ordinal scale includes the essential properties of a nominal scale, plus two more:

1. The categories are mutually exclusive.
2. They are ordered according to the amount of the attribute they represent.

An interval scale is characterized by three basic properties, including those of nominal and ordinal data.

1. The categories are mutually exclusive.
2. They are ordered according to the amount of the attribute they represent.

3. Equal differences in the attribute are represented by equal differences in the number assigned to the categories.
4. Numbers assigned to the categories are proportional to the amounts of the attribute represented by them.

A ratio scale has all of the attributes of an interval scale with an added difference: both interval and ratio levels are quantitative measurement. The difference between an interval and a ratio scale is that the interval scale has an *arbitrary* zero point, while the ratio scale has an *absolute* zero point.

The use of quantitative and/or qualitative measures is also an important consideration when planning for the data collection phase of critical analysis. Quantitative measures are typically numerical, and qualitative measures are textual. Qualitative variables are aggregate indicators of the magnitude of concepts, whereas qualitative variables show the character or content of concepts. It is important to keep in mind that neither quantitative nor qualitative variables are better or worse. The use of either type of data should be consistent with the research questions that one is asking, and should not be determined because one has a preference for statistical or qualitative analysis.

Three kinds of "averages" are defined from data that have been obtained for research questions. The *mean* is the point around which the values in the distribution balance; it is the mathematical or arithmetic average. In order to calculate a mean, at least internal-level data exist. The *median* gives information about the value of the middle position in the distribution. It is the point in the distribution of values at which 50 percent of the scores fall below and 50 percent of the scores fall above. In order to calculate a median, you must have at least an ordinally measured variable. *Mode* represents the most frequent value in distribution. The mode is the simplest measure of averages and is, therefore, not viewed as an overly precise or informative measure of average. In addition, the mode is the only measure that is appropriate for nominal data.

As mentioned at the beginning of this chapter, every organization must have a road map or guide as to where it wants to go or believes that it is going. This must start by knowing the organization's past, present, and desired future. This necessitates having a plan. This chapter has shown through classical studies how these theoretical principles are also true for occupational safety management.

EXERCISE

Maslow's hierarchy of needs is a core competency that an occupational safety manager should understand. It can be used for the manager to understand how employees interact in real life, including the work environment. Explain how Maslow's hierarchy can be used to explain internal theft by employees. How would this explanation be different from using Maslow's theory to explain theft from outsiders?

REFERENCES

Herzberg, F., Mausner, B., and Snyderman, B. B. (1959). *The Motivation to Work*. New York: John Wiley.
Maslow, A. H. (1954). *Motivation and Personality*. New York: Harper and Row.

4 An Occupational Safety Management Program

Occupational safety management should take a comprehensive approach. Occupational safety management is a broad field of management related to loss prevention, physical and logical controls, occupational safety, and asset protection functions. It entails the identification of an organization's employees, physical assets, and intangible assets and the development, documentation, and implementation of physical measures, policies, procedures, and guidelines.

Occupational safety management is a process that is used to develop mechanisms to protect persons. Occupational safety management tools such as risk assessments, criticality and vulnerability studies, and cost–benefit analyses show just a glimpse of the methodological tools of the occupational safety manager. The risk assessment itself looks at the value of organizational assets, accounting for both the criticality and vulnerability of the asset to determine the correct amount of the occupational safety investment. The comprehensive process of risk analysis, as well as other occupational safety management tools, utilizes a process of critical thinking while providing a basis for a comprehensive occupational safety management program.

UNDERSTANDING THE ORGANIZATION TO ESTABLISH OCCUPATIONAL SAFETY PROGRAMS

The role of occupational safety management must begin by asking some fundamental questions to determine where to begin in the process of protecting organizational assets. The findings, provided by a critical thinking framework, are based on the analysis of several occupational safety–related factors, which serve as the basis for the implementation of occupational safety measures to develop an optimal level of protection. The desired level of protection is the degree of occupational safety provided by a particular countermeasure or set of countermeasures to protect the asset.

IDENTIFY ORGANIZATIONAL ASSETS AND GOALS

The first step of the occupational safety management planning process is to identify organizational assets and specific overall goals for the organization. This part of the planning process should include a detailed overview of each organizational asset and its relationship to organizational goals, including the reason for its selection and the anticipated outcomes of goal-related projects. Where possible, objectives should be described in quantitative or qualitative terms. These quantitative and qualitative goals should be measurable. Being able to measure needs, as well as outcomes, is fundamental to occupational safety management and to the entire organization.

The application of occupational safety programs should follow the critical thinking process regardless of the type of asset involved. The asset has a direct impact on the application of this process and how countermeasure recommendations resulting from the process are to be implemented. Application of this occupational safety standard ensures a comprehensive approach to meeting organizational occupational safety needs in the threat environment, and ensures that the scope of occupational safety is commensurate with the risk posed to an asset relative to cost.

Each goal should have financial and human resources considerations associated with its achievement. As an example, suppose that we have an organization that manufactures widgets, and if the organization has a goal to increase production and sales by 20%, how does that impact the role of occupational safety? What aspects of organization assets, risks, and vulnerabilities change when the overall goals of the organization change? Maybe assets, risks, and vulnerabilities don't change. However, we do not know that until we adequately understand the organizational environment.

Therefore, a primary goal in occupational safety management is to identify and evaluate organizational assets. For simplicity, we can categorize assets as people, physical assets, and intangible assets.

People

The people category can range from the line workers doing work of minimal strategic importance to vital individuals who hold key roles and whose incapacity or absence will affect the business.

Physical Assets

This includes all physical organizational assets: real estate, buildings, facilities, equipment, materials, monies, supplies, inventories, and all other physical resources that allow the organization to operate.

Intangible Assets

The intangible assets category includes information, plans, and organizational strategies. It can be marketing and sales plans, detailed financial data, trade secrets, personnel information, sensitive office correspondence, and minutes of meetings. It can also include things such as a positive public perception of an organization. Often, these organizational assets are overlooked at first glance. However, the occupational safety manager as well as organizational leadership should work together to illuminate and assess the significance of these assets.

The Relationship between Organizational Assets

The occupational safety manager must understand the relationship of assets in organizational structures and functions. This includes the physical and logical relationships that assets have with each other in the organizational environment. It can also relate to the impact that one asset can have on others in a loss event. An essential tool for better understanding the relationship among organizational assets is through the configuration of an asset hierarchy. A very simplistic asset hierarchy diagram for our widget-making machine may look like this:

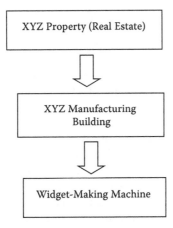

From this illustration, we must assume that a loss event impacting XYZ Property (real estate) may impact the building and, subsequently, the widget-making machine. For example, if the property floods, that may result in the building flooding, which in turn may result in loss to the widget-making machine. In another example, a loss to the XYZ Manufacturing Building asset, such as a fire, may also have an impact on the widget-making machine. Or, in a final situation, an employee could misuse the widget-making machine so as to make it inoperable, resulting in loss to the organization. So from these various loss scenarios, we can see that there can be a relationship where one asset loss is attributed and related to other losses in a hierarchical relationship that the asset shares with other organizational assets.

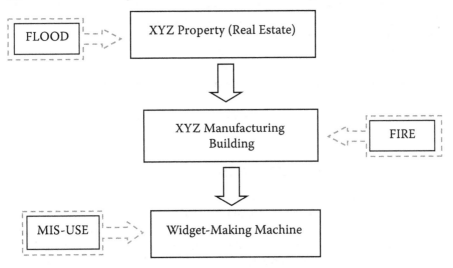

The point of the asset hierarchy is not as much about making a rigid model of what the assets are or where they reside in the organizational structure, but rather to enlist the asset hierarchy as a critical thinking tool to better understand the physical and logical relationships that assets have with each other in the organizational environment and the impact that one asset can have on the other in a loss event.

IDENTIFY THE CRITICALITY OF ASSETS IN MEETING ORGANIZATIONAL GOALS

Determine the asset's criticality for the organization. This evaluation is where individual assets are linked with the organization's strategic plan. In the organizational environment, the occupational safety manager must understand the value (or criticality) of an asset in attaining organizational goals. If the organizational goal is to increase widget production by 20 percent in the following year, what is the role of the assets in goal attainment?

A simplistic way to look at the criticality of organizational assets is to place a criticality value on that asset to provide a numeric valuation system that expresses the value of the asset to the organization. For example, we could lump assets into three general categories: critical, not critical, and no criticality value. From these three general views, the occupational safety manager can also include redundancy criteria in the asset criticality measurement. From this scenario, the occupational safety manager can rate organizational assets on a scale of 1 to 5 for criticality, with 5 being the most critical (with no redundancy) and 1 having no criticality in meeting organizational goals. The scale is as follows:

5: Critical asset with no redundancy
4: Critical asset with redundancy
3: Not critical asset with no redundancy
2: Not critical asset with redundancy
1: An asset with no criticality to the organization

Therefore, an asset that is critical to the operation and has no onsite replacement would be given a criticality of 5, one that is critical to the operation and has a replacement onsite would be given a criticality of 4, and so on.

The premise of understanding organizational assets and their relationship with meeting organizational goals is crucial because failure to fully understand organizational assets will result in the over- or underprotection of those assets. Overprotection of assets will cost the occupational safety manager resources that may be better used elsewhere, whereas underprotection of assets can contribute to greater exposure to risk and subsequent loss.

EXISTING LEVELS OF ASSET PROTECTION

Once all organizational assets are evaluated, the next step is to determine their existing level of occupational safety protection. Depending on the organization's requirements, assets may be classified into two or more levels of occupational safety need. The occupational safety manager should be cautious of having too many existing levels of occupational safety protection; this tends to dilute their importance, and it overcomplicates occupational safety management. Having too many occupational safety levels also proves expensive in terms of employee training and occupational safety resources and practices, in that the costs are often greater than the potential losses. Keep it simple by providing adequate protection while minimizing complexity in its application.

The occupational safety manager can utilize the concept of levels of protection to gain an understanding of occupational safety needs by forming needs into three general categories:

1. *Existing level of protection*: the degree of occupational safety provided by countermeasures to an asset currently being utilized at an organization
2. *Necessary level of protection*: the degree of occupational safety provided by a set of identified countermeasures, which must be implemented and justified by a risk assessment to provide a necessary level of protection
3. *Optimal level of protection*: the degree of occupational safety needed to completely mitigate organizational risks

The existing level of protection may be determined through site surveys, interviews, reviews of policies and procedures, testing, and so on to determine what countermeasures are currently in place and how effective they are. Current conditions may then be matched up against the existing level of protection and then compared to the necessary level of protection to determine if the current level adequately addresses the threat(s) or if vulnerabilities need to be addressed.

In determining existing levels of protection, the occupational safety manager should define in detail the following key areas of existing occupational safety management:

- Existing asset classification practices: guidelines for specifying occupational safety
- Previous risk assessments and organizational understanding of an asset's risk
- Assignment of organizational asset ownership: is there an assignment of roles for handling organizational assets?
- Existing asset responsibilities: the tasks and procedures to be followed by the entities handling the asset
- Existing policies regarding mishandling of organizational assets, including how occupational safety violations are reported and dealt with
- Existing occupational safety awareness practices: education programs and classification of assets
- Existing occupational safety audit procedures: unannounced checks of occupational safety measures put in place to find out whether they are functioning

If the existing level of protection equates to the necessary level of protection, current countermeasures should be maintained and evaluated on a regular basis. However, organizational conditions should be monitored for changes that may impact the effectiveness of countermeasures or the needed level of protection. Therefore, if the existing level of protection does not sufficiently address the risks, shortfalls must be identified and countermeasures to address those vulnerabilities must be considered for implementation.

ANALYZING RISKS TO ORGANIZATIONAL ASSETS

An effective occupational safety management system demonstrates a careful evaluation of how much occupational safety is needed to protect organizational assets. Occupational safety managers must realize that too little occupational safety means that organizational assets can easily be compromised, while too much occupational safety can make assets hard to use or so degraded that performance is negatively affected. Occupational safety must be inversely proportional to an asset's utility. It is given that there is always going to be risk associated with assets and activities. The only way to completely eliminate the risk would be, in many cases, to make that asset inoperable. Therefore, the role of occupational safety management is to find the optimal relationship between organizational processes, assets, and functionality. Although the risk assessment process will be more deeply analyzed in Chapter 7, it still maintains merit when looking at its overall role in building an occupational safety management program.

If the organizational asset is viewed through the inventory of potential loss events, the occupational safety manager must recognize that findings are not necessarily all inclusive. For each undesirable event where the assessed risk is either less than or exceeds the baseline level of protection, the occupational safety manager must identify the countermeasures that will provide a level of protection equivalent to the level of risk. For lower lever risk, minimum countermeasures are typically less stringent, but they may also be less effective in mitigating higher risks, while at the other extreme very high countermeasures are typically more stringent and generally more effective.

A minimum level of risk should be mitigated by minimal countermeasures, a low level of risk should be mitigated by low-level countermeasures, and so on. By adjusting from the baseline level of protection to determine which countermeasures are applicable to the assessed risks and identify changes from the baseline level of protection, the necessary level of protection can be determined. Once the level of protection necessary to meet the risk is identified, an evaluation of current conditions must be made to identify the existing countermeasures.

Occupational safety managers must understand that some degree of risk is assumed with all organizational activities and assets. Having no risk is virtually impossible, even with unlimited resources. Therefore, identify all risks to organizational assets, then determine the risks to accept and others to mitigate via occupational safety measures. The occupational safety manager must work with others throughout the organization to understand the effect on the business if the asset is lost or compromised. By doing this, you get a good idea of how resources should be assigned to protecting the asset.

For occupational safety managers, risk management is a fundamental component of their position. Risk management is a comprehensive approach to allocating resources for the protection of assets to achieve an acceptable relationship between risk and protection. Risk management decisions are based on the application of risk assessment, risk mitigation, and risk acceptance. The risk assessment itself looks at the value of organizational assets, accounting for both the criticality and vulnerability of the asset to determine the correct amount of occupational safety investment.

The comprehensive process of risk analysis utilizes a process of critical thinking. The risk assessments will be the basis for management of risk through the application of strategies and countermeasures to reduce the threat of, vulnerability to, and/or consequences from an undesirable event. Risk acceptance is an explicit or implicit assumption that some risk is not feasibly promising to mitigate.

The desired levels of protection of assets should be critically determined via a risk-based analytical process or risk assessment. The process will focus on risk as a measurement of potential harm or loss from an undesirable event. Understanding risk means understanding threats, vulnerabilities, and consequences. The level of risk is the combined measure of threats, vulnerabilities, and consequences posed to assets from specified loss events.

If the existing level of protection is insufficient, a determination must be made as to whether the necessary level of protection can be achieved; specifically, if the countermeasure can be physically implemented and whether the investment is cost effective. Cost effectiveness is based on the investment in the countermeasure versus the value of the asset. In some cases, investment in an expensive countermeasure may not be advisable because the life cycle of the asset is almost expired. Additionally, consideration should be given to whether other countermeasures may take priority for funding. Note that *cost effective* is a different value from *cost prohibitive*. A countermeasure is cost prohibitive if its cost exceeds available funding. Funding may exist for a countermeasure, but it may not be a sound financial decision to expend that money for little gain, making the countermeasure not cost effective.

There are a variety of mathematical models available to calculate risk and to illustrate the impact of increasing protective measures on organizational assets. For the purposes of this standard, an assumption is made at this step of the process that there are no countermeasures in place and that complete vulnerability exists. This approach is necessary to ensure that all occupational safety criteria will be considered as the process is completed, and to define the relationship between the level of risk and the level of protection. The level of risk must be mitigated by a commensurate level of protection. Often, a high level of risk must be mitigated by implementing a high level of protection.

Risk assessments utilize a form of critical thinking. Glaser's (1941, pp. 5–6) premise that critical thinking is (1) an attitude of being disposed to consider in a thoughtful way the problems and subjects that come within the range of one's experiences, (2) knowledge of the methods of logical inquiry and reasoning, and (3) some skill in applying those methods is very applicable in conducting risk assessments.

A very simplistic example of a risk assessment, utilizing critical thinking, could be evaluated by taking the asset value, multiplying it by the criticality of the asset, and then multiplying it by the vulnerability of the asset to determine an actual risk value. The equation is exemplified as $AV \times C \times V = RV$. To make this equation operational, the occupational safety manager can first take an asset and determine its value, then multiple it by its criticality value, which is represented as a percentage from 100% to 0% (or, in number form, from 0.0 to 1.0). Next, multiply by the vulnerability, which is also a percentage expressing the likelihood that a risk event will materialize into a loss.

The following is an example of a critical thinking risk assessment, or AV × C × V = RV:

Asset value (AV): AV considers an asset's value as the original cost, adjusted upward for improvements since the purchase of the asset and downward for loss in value related to the aging of the asset (Damodaran, 2012).

Asset criticality (C): C in the equation represents the value or degree of importance of an asset in regard to achieving organizational goals. For this value, the occupational safety manager can apply asset criticality as a percentage.

5: Critical asset with no redundancy—100%, or 1.0
4: Critical asset with redundancy—75%, or 0.75
3: Not critical asset with no redundancy—50%, or 0.5
2: Not critical asset with redundancy—25%, or 0.25
1: An asset with no criticality to the organization—0%, or 0.0

Asset vulnerability (V): V is the likelihood that a risk or threat event will occur to an organizational asset. Vulnerability is based on past loss data. Past loss data may be self-collected, or it can be collected from historical statistics from the organization's archived data, or from data from like organizations. The asset vulnerability would be defined as a percentage of likelihood that a specific loss event will occur in the budget period.

Risk value (RV): RV is the estimated cost of a risk to an organizational asset. It is determined by multiplying an asset's value by the asset's criticality to the organization, multiplied by the likelihood (vulnerability) of the occurrence of the risk event, resulting in the valuation of the risk in monetary values.

The following is an example of applying the risk value model (RVM) to a scenario as an aid to demonstrate risk value to an organizational asset. For simplicity's sake, the asset will be our widget-making machine, which costs our organization $10,000. The widget-making machine is critical to our organization, but we do have redundancy in that we always keep an extra machine on hand. Based on past loss data, the occupational safety manager knows that twice in the past 10 years, the widget-making machine was stolen and not recovered.

AV: Our asset value for the widget-making machine considers an asset's value as the original cost, adjusted downward for loss in value because it is 2 years old and has a life expectancy of 10 years. Therefore, AV = $10,000 × 0.80 (remaining life expectancy) or $8,000.

C: The widget-making machine is deemed a critical asset, but we do have redundancy. Therefore, 0.75 represents the value or degree of importance of the widget-making machine in achieving organizational goals.

V: The likelihood that a theft will occur to the widget-making machine is 20% annually. This assumption is based on past loss data, in that the occupational safety manager knows two of the machines were lost to theft in the last 10 years.

RV: From this exercise, the annual risk value of theft to the widget-making machine is as follows: $8,000 (AV) × 0.75 (C) × 0.20 (V) = $1,200.

So does that mean we can spend $1,200 annually to prevent the machine from being stolen? No, not necessarily. What it does mean is that we have utilized a critical thinking framework to give us a better understanding of our loss environment to an organizational asset. Based on this exercise, we can fiscally justify a system to protect the machine against theft. The occupational safety system to protect the machine may include bolting it to the floor, using measures to permanently identify it as the property of XYZ Company, assigning ownership and responsibility of the machine to the unit or staff, or directing policies and rules facilitating an increase in occupational safety to the widget-making machine.

The idea in the risk value model is to provide the occupational safety manager with a critical thinking tool to protect organizational assets. In practice, various risk assessment methodologies will provide varying outputs, from numbers and percentages to qualitative ratings. The occupational safety manager must determine what outputs from their respective methodologies correlate with the desired level of protection.

In an organization where multiple risk assessments may be conducted, the occupational safety manager will need to evaluate the comprehensive findings and determine what countermeasure recommendations to implement, or if a single risk assessment will be accepted for application. In gauging the value of a risk assessment, they should meet the following criteria at a minimum:

- The methodology must be credible, and it must assess the threat, consequences, and asset vulnerability to a specific loss event.
- The methodology must be reproducible, and it must produce similar or identical results when applied by various occupational safety professionals.
- The methodology must be defensible, and it must provide sufficient justification for deviation from the baseline.

For the occupational safety manager, the risk assessment is a fundamental component of determining protection efforts. The risk assessment can be a critical thinking tool utilized in allocating resources for the protection of assets. For the occupational safety manager, risk management decisions are based on the application of risk assessment.

COSTS VERSUS BENEFITS OF OCCUPATIONAL SAFETY

Cost considerations can be a primary factor in a decision to implement a countermeasure. A cost–benefit analysis is a critical thinking tool used by occupational safety managers in justifying occupational safety measures or programs. Cost–benefit analysis is a cost analysis methodology used to justify occupational safety expenditure; however, all costs, including life cycle costs, should be considered in whatever methodology is used. In addition to direct project costs, those expenditures associated with indirect impacts (e.g., business interruption, loss in productivity, or loss in credibility) should be considered. Any decision to not secure assets or to defer implementation of occupational safety measures due to cost (or other factors) must be documented, including the acceptance or transfer of risk.

The cost–benefit analysis is a comparative assessment of the benefits from your occupational safety measure and the costs to perform it, in relation to the financial impact resulting from potential loss to the asset. In a cost–benefit analysis, everything gets a dollar value. The occupational safety manager should consider the costs for all phases of the occupational safety project. Costs may be one-time capital or recurring costs such as personnel time, supplies, materials, or maintenance. The determining criterion in a cost–benefit analysis is that the occupational safety benefit exceeds the cost. How much the benefit exceeds the cost in determining occupational safety measure implementation is going to be based on the occupational safety manager in consultation with organizational leadership. The premise is that the cost–benefit analysis is another tool used by occupational safety management to critically deal with risk to assets.

IMPLEMENTING OCCUPATIONAL SAFETY PROGRAMS

When occupational safety managers complete the critical analysis via asset criticality and vulnerability studies, risk assessments, and cost–benefit analysis, they may implement the occupational safety measures that have been determined to best fit the asset risk. Implementation of new occupational safety programs is best accomplished through stages to make it easier for the organization to adapt to changes in the working environment. The occupational safety manager and organizational management should understand that there may be user resistance to occupational safety functions. It is recommended that staged implementation be performed starting with the most critical or vulnerable assets.

ASSIGNMENTS AND TIMELINES

As the organization implements occupational safety measures, it must establish timelines for completing the associated tasks. This portion of the occupational safety implementation process should consider the abilities of staff members and the time that is realistically necessary to complete projects. The amount of preparation required to implement occupational safety measures may limit their immediate achievability. If the occupational safety measure bears no capital cost, such as policy and procedural changes, or can be incorporated into a new project, the countermeasure can often be implemented immediately. When countermeasures require advance budgeting or coordination with outside vendors, implementation may be delayed.

MONITOR FOR COMPLIANCE

Effective occupational safety management depends on adequate compliance monitoring. Most often, violations of occupational safety practices, whether intentional or unintentional, become more frequent and serious if not detected and acted on. Compliance monitoring combines two primary activities: detecting occupational safety violations and responding to them.

The occupational safety manager should document the response to violations, and follow up immediately after noncompliance is detected. The organization should have a designated response group to deal with occupational safety violations. Members of the response group should have access to organizational leadership so that severe situations can be dealt with.

A critical part of noncompliance should be the generation of reports for organizational leadership that discuss occupational safety violations. An additional objective of monitoring occupational safety measures for noncompliance is to identify potential occupational safety violations before they dilute the effectiveness of the program or cause serious damage.

REEVALUATE ASSETS AND RISKS

Occupational safety management is a discipline that should be dynamic. As changes in the organization or assets occur, a reassessment of the occupational safety measures should also occur. Organizational leadership should keep occupational safety management abreast of larger changes in the organization so that occupational safety operations and measures are prepared to meet those challenges.

Even the best-laid plans can sometimes be changed by unanticipated events. An occupational safety management plan should include contingencies if certain aspects of the master plan prove to be unattainable. Alternative courses of action can be incorporated into each segment of the planning process, or for the plan in its entirety. The occupational safety manger must continually identify and analyze threats and vulnerabilities to assets, and recommend and implement appropriate countermeasures.

CONCLUSION

Developing an occupational safety management program requires a broad field of knowledge in asset loss prevention, physical occupational safety, occupational safety, and intangible asset protection functions. It requires a comprehensive knowledge of organizational assets and the development and implementation of physical measures, policies, procedures, and guidelines to protect those assets

Occupational safety management requires critical thinking skills in developing mechanisms to protect organizational assets. The process of occupational safety management utilizes processes of critical thinking, providing a basis for a comprehensive occupational safety management program. The dynamic nature of organizations and environments requires that the occupational safety response also be dynamic.

EXERCISE

Is occupational safety an expense or an investment? Compare and contrast the expense–investment relationship, and explain this in detail.

REFERENCES

Damodaran, A. (2012). *Investment Valuation: Tools and Techniques for Determining the Value of Any Asset.* 3rd ed. Hoboken, NJ: John Wiley.   ·
Glaser, E. M. (1941). *An Experiment in the Development of Critical Thinking.* New York: Teachers College, Columbia University.

5 Developing Policies and Procedures for Occupational Safety Management

Organizational vision and mission are further defined by policies and procedures. The occupational safety unit must develop policies and procedures to promote the organization's vision and mission while protecting organizational assets. A policy tells us what to do, while a procedure tells us how to do it. Many organizational policies can be written around laws, regulations, and standards.

The Occupational Safety and Health Administration (OSHA) applies to all employers that operate under federal government jurisdiction. Occupational safety coverage is provided either directly by federal OSHA or through an OSHA-approved state program, with the exception of self-employed persons, farms at which only immediate members of the farm employer's family are employed, and working environments regulated by other federal agencies under other federal statutes. Even when another federal agency is authorized to regulate safety, health, and working conditions in a particular industry, if it does not do so in specific areas, then OSHA standards apply.

OSHA provides the laws and legal definitions of how occupational safety should be addressed in an organization. While standards published by organizations such as the American National Standards Institute (ANSI) are "guides" for how you could address a hazard, ANSI standards look a lot like regulations. They're detailed, technical instructions for addressing hazards.

For this chapter, commonly used definitions are as follows:

Policy: the organization's guiding or governing principle
Procedure: tells people how to do something; may comprise several work instructions
Protocol: another name for procedure; associated with specific disciplines.
Process: a series of interrelated activities that result in an outcome; comprises several procedures
SOP: standard operating procedure
Work instruction: specific steps within a procedure; assumes one person or job completes the task from start to finish; can be used for training; usually unit contained or job specific

A policy is the organization's guiding or governing principle. It is a general guideline that sets forth parameters for decision making along with the authority for implementation. Policy development is a planned process requiring due diligence in thought as a policy becomes a strong principle of the organization. Planners must look to the desired outcomes while considering the needs of end users. Policy justification should reflect problem solving while demonstrating persuasive reasoning, clarity, and coherence. How will the policy be enforced? Is there an incentive (including negative incentives)? Is there support for the policy from the workforce? Is there some overriding need for the policy? The policy should be consistent with the organization's mission, culture, strategy, and vision. It should not overlap or contradict other policies and procedures.

Clarity is important in writing a policy. The average reader should not have any questions about the objective of the policy. When writing, do sufficient research to assure you are compliant with other policy and legal requirements. All stakeholders should have an opportunity to have input and give feedback about the policy. In a large organization, worker committees or teams may be organized to assure that feedback is received from all levels. When written, policies must be published in a form that is easily accessible to those affected by the policy. In general, a policy manual should be easy to use with a format that allows for ease in updating.

A policy or procedure should be written (or revised) for fairness, consistency, and accountability when there is confusion about the appropriate action to take in a given situation. Sometimes, legislative or regulatory changes require a change in policy or procedure. After a while, a policy or procedure may have so many exceptions, exemptions, or waivers that it is ineffective. This requires a review and either a rewrite or enforcement of the present policy (procedure).

Not all actions require or fall within a written policy. While policy sets out the order of business, anomalies occur and are handled on a case-by-case basis. Policy makes for efficient work. We may be tempted to write a policy for every possible event and become inefficient by spending valuable resources writing nothing but policy, and ending up with a policy manual so large it won't be read. The manual becomes an organ to catch people doing wrong rather than to direct people in doing right.

The policy should begin with a policy statement. The policy statement explains what you are doing. Is the policy a standard or a guideline? Standards are specific requirements that must be met, while guidelines identify best practices. The statement tells when the policy applies and lists major conditions or restrictions. Next is the reason for the policy and a description of the conflict or problem the policy is designed to resolve. List those who should know this policy and should follow the policy and its accompanying procedures in order to do their job. Document the resource and contact information at the end of the policy.

Procedures tell us how to do something; they are the steps for getting things done. Inside of a procedure may be work instruction: specific steps within the procedure. Work instructions usually involve one person accomplishing the task. The aim for a procedure is to use one action per step and to assign the action to whoever is responsible for the action. Many general considerations of procedures occur when writing policies. In lieu of repeating material, just keep in mind those ideas presented in this chapter.

A procedure begins with a purpose statement. What are you trying to accomplish? Then list the actions and sequence, who does each step (to include handing off to another person), where and when this must be done, and any standards for completing the work.

When a new policy or procedure is adopted, the work population must be educated as to its existence. In the case of a procedure, this may require training. Whether through a training session or a briefing, notice should be formal, with documentation of all persons receiving the new material. New employees should be notified of policies and procedures in new employee orientations.

When the policy or procedure is in effect, enforcement is usually through disciplinary action. Discipline must be consistent, or the policy or procedure will lose its power. Depending on the severity of a policy violation, progressive discipline provides a level of punishment commensurate with the violation. For example, a first offense may result in a one-day suspension, a second offense in a one-week suspension, and a third violation in termination. In such a case, two employees could get a different punishment for the same conduct. Another consideration may be the seniority of employees. Discipline may be more severe for a seasoned veteran than for a new employee. Some policy violations are so severe as to warrant immediate termination. These are usually reserved for those violations that could result in death or serious injury to others, or those relating to matters of integrity. Embezzling money from the organization may not create a risk of injury, but the nature of the violation would warrant dismissal.

When a policy or procedure has not been enforced, or has been selectively enforced, reviving the policy or procedure follows basically the same notification process as with a new or rewritten one. When reviving a policy or procedure, personnel are formally trained with the additional information that the policy or procedure has not been enforced, that it is still applicable and necessary to the function of the organization, and that, forthwith, it will be enforced. Consideration should be given to a graduated disciplinary process in reimplementing the policy or procedure.

Policies affecting the workforce are as varied as the imagination allows. Some of the most common are policies affecting access control, company equipment, visitors to the workplace, and other regulatory compliance. Policies affecting the general workforce are often seen as restrictive and a hassle. It is important for a policy to state its reason and for the policy to be applied throughout the organization. If the CEO doesn't have to comply, employees begin to question why they have to comply. Executives should set the example by following all policies and directing others to do so.

An occupational safety management system at the organizational level requires compliance with multiple standards, as well as a comprehensive safety approach to create a safe working environment for employees. For the organization, occupational safety requires compliance to multiple laws and regulations that have been determined to make the work environment a safer place. At the organizational level, the integration of an occupational safety management system should influence overall policy and management arrangements, and help to stress the importance of having a safe environment. From this perspective, occupational safety is a management responsibility that is organization-wide, and it should not be a mere task for departments and/or specialists.

Occupational safety should be woven throughout organizational policy. This will assist occupational safety and health (OSH) management by confirming organizational investment in occupational safety, proliferating a safety culture throughout the workforce, and simply making the work environment safer. Organizational policies should contain elements of OSH policy and staff participation.

OCCUPATIONAL SAFETY POLICIES
THAT AFFECT THE GENERAL WORKFORCE

ACCESS CONTROL

A policy for site access control should state reasons for buildings and doors to be open or locked and the time at which doors will be open or locked.

> The administration building doors will be unlocked Monday–Friday at 7 a.m. Access at other times is by key, or contact the safety department.

The policy should dictate who is entitled to a key, and a log should be maintained of all keys issued. The policy should require human resources to withhold final pay until keys are returned; for example:

> Keys are issued to all full-time employees for their office. Keys shall not be duplicated and are to remain in the possession of the employee at all times. Keys shall be returned upon resignation or termination; for example, supervisors are authorized a master key for all offices in their area. Master keys shall not be loaned out and shall remain in the possession of the supervisor at all times.

Electronic access control systems should dictate the times and days at which personnel are allowed to enter the premises. Access control cards and fobs shall be returned on resignation or termination. Employees should be aware that their card may be deactivated for security or safety reasons; for example:

> Employees are issued access control badges that shall be displayed on the person at all times while on the property. Badges should be visible above the waist by hanging them on a lanyard around the neck or by clipping them to a shirt, blouse, or jacket. Employees are not to loan the badge to another person.... Misuse can result in having your badge deactivated.

VISITORS

Visitor control policies should state the process for permitting visitors access to the facility and who has permission to sponsor a visitor; for example:

Any employee may sponsor a visitor to the facility. Visitors must be met at the front desk by the employee, who will sign for the visitor badge and escort the visitor at all times while they are in the facility. The employee is responsible for signing out the visitor when they leave the facility.

Visitor policies may also specify areas where visitors are not allowed access:

No visitors are allowed in the research lab at any time without the written approval of the lab director.

A visitor policy may contain a procedure that employees will follow if they find an unidentified visitor:

Any employee who observes an unescorted visitor or unknown person not displaying an employee badge shall immediately notify security and provide a description of the person. The employee is not to approach the unknown person.

COMPANY EQUIPMENT

A policy for equipment control should state the authorizations needed to remove property from the premises:

All company equipment is marked with an identification tag and must be checked out before leaving the facility. The employee must have the approval of a supervisor to remove property from the facility.

CODE OF ETHICS

The occupational safety force should adopt or develop a code of ethics. All personnel should receive training on the code of ethics and agree to abide by the code. For example:

All occupational safety personnel will receive _ hours of ethics training. At the completion of the training, each employee will sign an agreement to abide by the code of ethics.

USE OF EQUIPMENT

A directive should govern the use of equipment for work purposes only:

All property issued to the employee is for official use only and is intended to be used only in the line of duty when circumstances warrant their use.

OSHA Recordkeeping and Report Writing

A policy on reports should include when a report should be written and what should be included in the report. For example:

> Occupational safety reports must be completed when:
>
> 1. There is a violation of an occupational safety order or safety rule.
> 2. Someone reports an incident to occupational safety.

Furthermore, compliance to OSHA requires that employers with more than ten employees and whose establishments are not classified as a partially exempt industry must record work-related injuries and illnesses using OSHA Forms 300, 300A, and 301. Partially exempt industries include establishments in specific low-hazard industries such as retail, service, finance, insurance, and real estate.

Employers who are required to keep Form 300, the "Injury and Illness" log, must post Form 300A, the "Summary of Work-Related Injuries and Illnesses," in a workplace every year from February 1 to April 30. Current and former employees, or their representatives, have the right to access injury and illness records. Employers must give the requester a copy of the relevant record(s) by the end of the next business day.

CONCLUSION

A policy is the organization's guiding or governing principle. It has been the intent of this chapter to assure that general guidelines be set forth with parameters for decision making along with the authority for implementation. Policy development should be a planned process requiring due diligence in thought as a policy becomes a strong principle of the organization. Policy justification should reflect problem solving while demonstrating persuasive reasoning, clarity, and coherence.

EXERCISE

It has come to the attention of occupational safety management that your organization is having issues with employees leaving work early, with another employee clocking out for them later. Write a policy that will deal with this situation. How will the policy be enforced? Is there an incentive (including negative incentives)? Is there support for the policy from the workforce? Is there some overriding need for the policy? The policy should be consistent with the organization's mission, culture, strategy, and vision. It should not overlap or contradict other policies and procedures.

6 Staffing and Occupational Safety Management

The area of staffing is complex for the occupational safety manager. The roles of safety and staffing are complex because of the varying characteristics of the workforce. First, you have the entire staff of an organization (with differing skill sets and levels of understanding) as well as temporary and contract employees to consider in terms of occupation safety management. Each of these groups will require differing levels of understanding and guidance in creating the optimal safety environment for an organization. Temporary workers are employed in some of the most hazardous jobs, including waste recycling, fish processing, and construction. With little information about the nature of their jobs and an absence of safety training, workers report retaliation for health and safety complaints along with an increased risk of work-related injuries and illnesses (Freeman and Gonos, 2011).

Safety management for staffing is more than just creating policies and procedures for dealing with hazards and risks. It deals with the manner in which safety is handled in the workplace and how those policies and procedures are implemented by the workforce. Kennedy and Kirwan (1998) assert that the nature by which safety is managed in the workplace (e.g., resources, policies, practices and procedures, and monitoring) will be influenced by the safety culture or climate of the organization. An organization's safety culture is ultimately reflected in the way that safety is managed in the workplace.

The intent of this chapter is to show several relationships between safety management and staffing as a means to provide a more safety-aware and engaged workforce. This chapter makes the assumption that employees have a skill set that has prepared them to adequately perform the requirements of their position in the organization. The role of safety management is not to train someone how to do a job, but rather to assist it in a way that will make tasks and jobs optimally correlate between safety and production. This chapter will address training, personal and professional codes of conduct, and the fostering of a culture promoting a safe work environment.

For temporary employees, staffing agencies hire, provide salary and benefits, and train the staff. Your organization will have someone who will be the liaison with the contract agency. In some matters, the staffing company may be liable for the intentional or negligent acts of staff. The liaison will be responsible for assuring that the contract agency is performing according to the specifications in the contract and meeting with agency representatives for routine performance reviews.

In some organizations, staff may be represented by a union and operate under a collective bargaining agreement with management. The safety manager must work with human resources and legal counsel to assure that discipline, scheduling, and promotions are consistent with the agreement.

Regardless of whether you have proprietary or contract staff, a job assessment will identify the needs for the staff force and begin the process for writing a job description. To write a valid job description, begin with listing all of the tasks the staff member will perform, such as answering the telephone, writing reports, watching monitors, or operating machinery. Any movements should be documented to justify requirements of the Americans with Disabilities Act (ADA). The ADA protects against discrimination of a person who can perform the job functions with or without reasonable accommodation. For example, list tasks such as standing, walking, negotiating stairs, climbing, crawling, reaching, and so on. Vision and hearing requirements, as well as the ability to function specific machinery, should be included in the list. As you screen applications for employment, those who cannot perform the functions with or without reasonable accommodation can be excluded from consideration. Is it important to write a thorough job description so potential applicants will know the requirements and apply for jobs for which they are qualified.

TRAINING

Staff must be provided with an understanding of safe practices for all of their functional occupational roles in the organization. Organizations are responsible for hiring, providing salary and benefits to, and training the staff. Liability rests solely on your organization for intentional or negligent acts of individual staff members. In addition to the regular employed staff, many organizations use contract employees who are managed via a contract company (often a company specializing in staff) for a specified contract price.

The job description is also a key to training. Training assures that the employee can perform job tasks proficiently. It minimizes civil damages in the event of a lawsuit. Training may be in the classroom, online, on the job, or via a combination of methods. Contract agencies provide a level of training for uniformity of their staff force. A proprietary staff force may be trained onsite or sent to a school that provides a basic level of training. Private corporations and safety organizations have ready-made materials for use in the general training of their staff force. Depending on specific needs, other prepared training programs may be helpful and even necessary. For example, a staff at a casino where alcohol is served may have required training from the state gaming commission as well as the alcohol beverage commission.

Staff should be aware of the company's policies and procedures, operational needs, and priorities. Orientation should include a copy of the policy manual. After a reasonable time, have the staff sign a form indicating they have read the policy manual.

Training should include general duties applicable to the staff force as well as specific duties for each post at which the individual staff member may work. Training should be conducted based on the frequency or criticality of a specific function. For example, reports completed daily have a high frequency, so training on report writing would be a reasonable training need. Use of force may not be frequent, but it is highly critical as the improper use of force can lead to injury and possible lawsuits. Training about force parameters will assure the proper conduct of staff and mitigate damages if and when force is used.

Other training needs would be patrol or station procedures, customer service, telephone communications, and other skills not necessarily related specifically to the trade, but for events that could impact the organization. For example, when staff answer the telephone or give instructions to visitors, they must sound (and appear) professional. Answering the phone with "Yeah, what can I do for you?" is quite informal and may present a negative impression of the organization. Proper procedures should be identified and taught to all personnel.

Training should include all equipment the staff will use. Many times, staff members are the first responders to a fire, safety, or medical emergency, so they should have diverse safety training. Generally, knowledge and use of fire extinguishers and other firefighting equipment are essential. The staff should know when and how to deploy and use safety equipment, such as personal protective equipment, a fall protection safety harness, or safety barriers. Training should include Occupational Safety and Health Administration (OSHA) rules that apply to the facility and specifically to activities the staff may encounter. All staff should be familiar with any first aid equipment and automated external defibrillators, if available. Staff should have a working knowledge of emergency evacuation and shelter-in-place procedures. The staff force should always participate in all drills and exercises. Most organizations expect the staff to coordinate emergency activities.

PERSONAL AND PROFESSIONAL CODES OF CONDUCT

Staff should be encouraged to develop personal and professional codes of conduct in the organization. Just like the Board of Certified Safety Professionals (BCSP) recognizes a code of ethics and professional conduct for its members, likewise all staff members of an organization should be inspired to develop their own personal and professional codes of safe conduct. Their code should detail ethics and professional standards in a way that is similar to the BCSP's code.

The BCSP, in their safety activities, should always sustain and advance the integrity, honor, and prestige of the safety profession by adhering to ethical and professional standards. These standards are shown as an example to be applied to all organizational staff, as well as staff directly involved in the safety department. Those standards are as follows:

1. Hold paramount the safety and health of people, the protection of the environment, and the protection of property in the performance of professional duties, and exercise the obligation to advise employers, clients, employees, the public, and appropriate authorities of danger and unacceptable risks to people, the environment, or property.
2. Be honest, fair, and impartial; act with responsibility and integrity. Adhere to high standards of ethical conduct with balanced care for the interests of the public, employers, clients, employees, colleagues, and the profession. Avoid all conduct or practice that is likely to discredit the profession or deceive the public.

3. Issue public statements only in an objective and truthful manner and only when founded upon knowledge of the facts and competence in the subject matter.

4. Undertake assignments only when qualified by education or experience in the specific technical fields involved. Accept responsibility for continuing professional development by acquiring and maintaining competence through continuing education, experience, and professional training.

5. Avoid deceptive acts that falsify or misrepresent academic or professional qualifications. Do not misrepresent or exaggerate the degree of responsibility in or for the subject matter of prior assignments. Presentations incident to the solicitation of employment shall not misrepresent pertinent facts concerning employers, employees, associates, or past accomplishments with the intent and purpose of enhancing qualifications and work.

6. Conduct their professional relations by the highest standards of integrity, and avoid compromise of their professional judgment by conflicts of interest.

7. Act in a manner free of bias with regard to religion, ethnicity, gender, age, national origin, sexual orientation, or disability.

8. Seek opportunities to be of constructive service in civic affairs, and work for the advancement of the safety, health, and well-being of their community and their profession by sharing their knowledge and skills.

The code of the BCSP details the ethical and professional standards that a safety manager would like to see in the workforce. If the safety manager, through training or the creating of a safety culture, instills into all staff the importance of developing personal and professional codes of conduct in the organization, the organization will prosper from it.

SAFETY CULTURE

Providing a safety culture is a key part of assuring a safe work environment. It is the task of the safety manager to assure that new hires as well as existing staff understand the basic elements of organizing, planning, and managing an effective safety program. The safety manager also establishes professional training programs for full-time and collateral-duty safety staff.

Cultivating a safety culture affirms the stance that safety should be a part of everyone's role in an organization. Burman and Evans (2008) recognize that organizational leadership is the key to creating a positive safety culture. From this perspective, the occupational safety manager should be the primary change agent in creating a positive safety culture. Broadbent (2007) has recognized the positive influences of transformational leadership, via occupational safety management, in furthering a safety culture. The concept was introduced to explain how a lack of knowledge and understanding about risk and safety by employees and organizations contributes to collateral losses during disasters.

Even a receptionist or clerical employee can assume the task of a safety specialist and provide the initial response to a hazardous event. Situations may arise resulting in an immediate need for staff to move from their normal tasks. Safety is not their forte, and relying on them for a safety necessity may not be the safety manager's obvious choice, but it may be the best choice in dealing with immediate loss events. The decision to train, as well as to encourage a safety culture, takes into consideration the volume of employee and visitor traffic, the value or sensitivity of the assets and resources, the availability of first responders to respond, and their response time, as well as other safety measures in place.

A worthy safety culture was identified by Pidgeon and O'Leary (1994) as having four factors: (1) senior management commitment to safety; (2) a shared care and concern for hazards, and solicitude for their impacts on people; (3) realistic and flexible norms and rules about hazards; and (4) a continual reflection upon practice through monitoring, analysis, and feedback systems.

Others have defined a safety culture as the product of individual and group values, attitudes, perceptions, competencies, and patterns of behavior that determine the commitment to, and the style and proficiency of, an organization's health and safety management (HSC, 1993, p. 23). Another widely used definition of safety culture, developed by the Advisory Committee on the Safety of Nuclear Installations, identifies it as the product of individual and group values, attitudes, perceptions, competencies, and patterns of behavior that determine the commitment to, and the style and proficiency of, an organization's health and safety management. Organizations with a positive safety culture are characterized by communications founded on mutual trust, shared perceptions of the importance of safety, and confidence in the efficacy of preventive measures.

An organization's culture regarding safety is cultivated by the safety manager through the staff's attitudes, while others emphasize safety culture being expressed through their behavior and work activities. The safety culture of an organization acts as a guide as to how employees will behave in the workplace. Their behavior, or their cultural diffusion of occupational safety, will be influenced or determined by what behaviors are rewarded and acceptable within the workplace. Clarke and Ward (2006) notes that the safety culture is observed not only within the premises and conditions of the machinery but also in the attitudes and behaviors of the employees toward safety. Therefore, safety culture is not about physical protective controls, but is an understanding of what is safe and unsafe behavior in an organization.

Developing a positive organizational safety culture is a critical factor influencing multiple aspects of staff performance and its relationship with occupational safety. Safety culture provides enduring valuing and prioritization of both worker and public safety by each member in every level of an organization. It determines the extent to which individuals and groups will commit to personal responsibility for occupational safety. The idea of developing a safety culture in an organization facilitates a perseverance, enhancement, and communication of occupational safety concerns. The definition combines key issues such as personal commitment, responsibility, communication, and learning in ways that are strongly influenced by occupational safety management to affect everyone in the organization.

The relationships between safety management and staffing provide an opportunity for a more safety-aware and engaged workforce. While employees should have a skill set that has prepared them to adequately perform the requirements of their position in the organization, the proactive development of occupational safety awareness should permeate their organizational role. From this perspective, the role of safety management is not to train staff how to do a job, but rather to assist them in a way that will lead to enhanced safety via training, personal and professional codes of staff, and the fostering of a culture promoting a safe work environment.

EXERCISE

An occupational safety manager should train and prepare all staff for their next promotion, regardless of whether they will stay in the same department or move to another. One philosophy is "The better they are and the more successful they are, the better I look and the more success I have." Now, this is not always easy. You have to be a really good manager to fully train staff. But consider this: if they move up the ranks, even if they pass you on the corporate ladder, if you did your job right, you will be well connected and have strong supporters in those individuals. Explain your thoughts on this.

REFERENCES

Broadbent, D. G. (2007). Transforming safety on the veldt. In: The 4th Annual SAFEmap Africa Competency Based Safety Symposium, South Africa, September 14.

Burman, R., & Evans, A. J. (2008). Target zero: A culture of safety. *Defence Aviation Safety Centre Journal*, 22–27.

Clarke, S., & Ward, K. (2006). The role of leader influence, tactics and safety climate in engaging employees' safety participation. *Risk Analysis*, 26, 1175–1186.

Freeman, H., and Gonos, G. (2011). The challenge of temporary work in twenty-first century labor markets: flexibility with fairness for the low-wage temporary workforce. Available May 27, 2013, at http://papers.ssrn.com/sol3/papers.cfm?abstract_id=1971222

Health and Safety Commission (HSC). (1993). Third Report: Organizing for Safety. ACSNI Study Group on Human Factors. London: HMSO.

Kennedy, R., & Kirwan, B. (1998). Development of a hazard and operability-based method for identifying safety management vulnerabilities in high risk systems. *Safety Science*, 30(3), 249–274.

Pidgeon, N. F. & O'Leary, M. (1994). Organizational safety culture: Implications for aviation practice. In N. Johnston, N. McDonald, & R. Fuller (eds). *Aviation Psychology in Practice*. Avebury, UK: Avebury Technical Publ.

7 Enterprise Risk Management and Occupational Safety

Many times when thinking about occupational safety, physical controls or personal protective equipment (PPE) comes to mind. Control mechanisms and PPE are the tools and equipment used to secure your assets or protect an employee, but is this occupational safety management? For the purposes of this text, occupational safety management is much larger than identifying a lockout–tagout procedure for your energized equipment. Occupational safety management is a comprehensive management process using critical thinking within an operational framework to implement and operate a system of safety throughout an organization.

An occupational safety management system at the organizational level requires compliance of multiple standards, as well as a comprehensive safety approach to create a safe working environment for employees. For the organization, occupational safety requires compliance to multiple laws and regulations that have been determined to make the work environment a safer place. At the organizational level, the integration of an occupational safety management system should influence overall policy and management arrangements, as well as stress the importance of having a safe environment. From this perspective, occupational safety is a management responsibility that is organization-wide, and should not be a mere task for departments and/or specialists.

SYSTEMS MANAGEMENT

The "systems" approach to organizational safety can be operationalized via several different approaches to management; however, none is simpler than Deming's cycle of Plan–Do–Check–Act (PDCA). This four-step management method is often used in business for the control and continuous improvement of processes (Deming, 1986).

Occupational safety's PDCA should be woven throughout organizational policy. This will assist occupational safety and health (OSH) management by confirming organizational investment in occupational safety and the proliferation of a safety culture throughout the workforce, and simply by making the work environment safer. Organizational policies should contain elements of OSH policy and staff participation.

Organizing occupational safety throughout the organization legitimizes responsibility, accountability, training, and communication, while contributing to a positive safety culture. It makes sure that the management structure is in place, as well

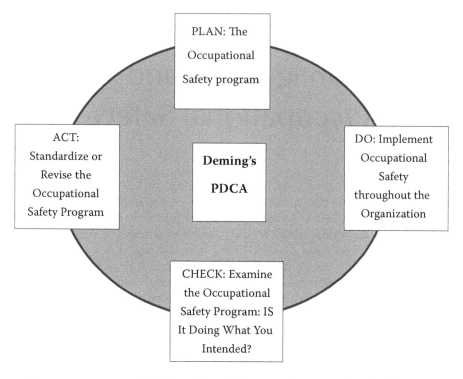

as the necessary responsibilities assigned for conveying occupational safety policy. Planning and implementation contain the elements of initial review, system planning, development and implementation, OSH objectives, and hazard prevention. Through the initial review, it shows where the organization stands concerning OSH, and it uses this as the baseline to implement the OSH policy.

Another "systems" approach to organizational safety can be instituted via Six Sigma's Define–Measure–Analyze–Improve–Control (DMAIC), which is another strategy for process improvement originally developed by Motorola in 1985. This five-step management method is often used in business for the control and continuous improvement of processes (Tennant, 2001).

While PDCA or DMAIC may define a context for program implementation, they also provide constructs to look at specific problems regarding occupational safety. These two models are not inclusive; there may be better models available for differing environments. The idea, as in critical thinking, is to follow a methodological process. Make continuous improvement through using a process of understanding the organizational safety environment.

APPLYING SAFETY STANDARDS TO A MANAGEMENT FRAMEWORK

After the occupational safety manager chooses a methodical framework to institute a comprehensive safety management program, the standards of safety that are required to make an organization "safe" can be applied. This is where safe-workplace laws

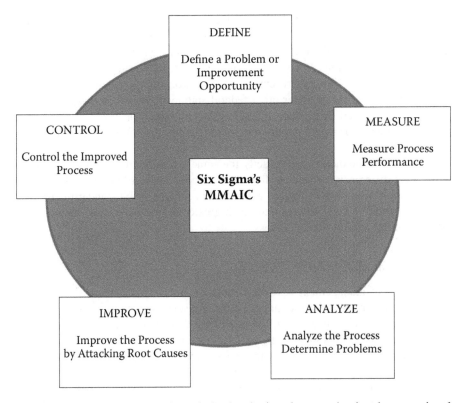

and regulations assume a major role in developing the organizational occupational safety program.

The Occupational Safety and Health Administration (OSHA) provides guidelines to reduce hazards to the workforce. OSHA was created within the U.S. Department of Labor to do the following:

- Encourage employers and employees to reduce workplace hazards and to implement new or improved existing safety and health programs.
- Provide for research in occupational safety and health to develop innovative ways of dealing with occupational safety and health problems.
- Establish "separate but dependent responsibilities and rights" for employers and employees for the achievement of better safety and health conditions.
- Maintain a reporting and recordkeeping system to monitor job-related injuries and illnesses.
- Establish training programs to increase the number and competence of occupational safety and health personnel.
- Develop mandatory job safety and health standards, and enforce them effectively.
- Provide for the development, analysis, evaluation, and approval of state occupational safety and health programs (OSHA, 1996).

As discussed in Chapter 5, OSHA applies to all employers that operate under federal government jurisdiction. Occupational safety coverage is provided either directly by federal OSHA or through an OSHA-approved state program, with the exceptions of self-employed persons, farms at which only immediate members of the farm employer's family are employed, and working environments regulated by other federal agencies under other federal statutes. Even when another federal agency is authorized to regulate the safety, health, and working conditions in a particular industry, if it does not do so in specific areas, then OSHA standards apply (OSHA, 1996).

Many ANSI standards cover exactly the same issues addressed in OSHA standards. One difference has to do with technical scope. OSHA laws typically set out only a general framework to guard against a hazard. An ANSI standard is usually consistent with the law but goes into much greater depth. It provides the technical details that the statutes do not address.

ENTERPRISE RISK MANAGEMENT

Enterprise risk management (ERM) in business includes the methods and processes used by organizations to manage risks and seize opportunities related to the achievement of their objectives. ERM provides a framework for risk management, which typically involves identifying particular events or circumstances relevant to the organization's objectives, assessing them in terms of likelihood and magnitude of impact, determining a response strategy, and monitoring progress.

A comprehensive approach to ERM that would be valuable to occupational safety at the organizational level is ISO 31000. ISO 31000 was published as a standard in 2009, and it provides a guide on the implementation of risk management. The utility of ISO 31000:2009 is in its validity for any public, private, or community enterprise, association, group, or individual. ISO 31000:2009 can be applied to any type of risk, whatever its nature.

The design and implementation of risk management plans and frameworks will need to take into account the varying needs of a specific organization, as well as its particular objectives, context, structure, operations, processes, functions, projects, products, services, assets, and specific practices employed. It is intended that ISO 31000:2009 be used to complement risk management processes in existing and future standards. It provides a common approach in support of standards dealing with specific risks and/or sectors and does not replace those standards.

Because ISO 31000:2009 can be used by any public, private, or community enterprise, association, group, or individual, it can be applied throughout the life of an organization and to a wide range of activities, including strategies and decisions, operations, processes, functions, projects, products, services, and assets. It can also be applied to any type of risk. Therefore, ISO 31000:2009 can provide an adaptive framework for any environment.

Although the standard provides generic guidelines, it is not intended to promote uniformity of risk management across organizations. It should also be noted that the standard is not intended for certification, regulatory, or contractual use. It does not provide specific criteria for identifying the need for risk analysis, nor does it specify the type of risk analysis method that is required for a particular application. This standard does not refer to all techniques and does not deal specifically with safety. It is a generic risk management standard, and any references to safety are purely of an informative nature.

CONCLUSION

An occupational safety management system at the organizational level requires compliance with multiple standards, as well as a comprehensive safety approach to create a safe working environment for employees. The systems approach should utilize a management control system such as Deming's PDCA or Six Sigma's DMAIC.

The idea of ERM and occupational safety is a key component of occupational safety management. ERM addresses risk from the organizational level. It involves commitment from all areas of the organization to safety. Implementing enterprise risk management requires consideration to several areas (Walker and Shenkir, 2008):

- Resolve to proactively manage risk, rather than react to it. Implementing ERM takes total commitment by management.
- Clarify the organization's risk philosophy. Organizations need to know their risk capacity in terms of people capability and capital. The board and management must come to an understanding of risk.
- Develop a strategy. Since risk relates to the events or actions that jeopardize achievement of the organization's objectives, effective risk management depends on an understanding of the organization's strategy and goals.
- Think broadly and examine carefully events that may affect the organiza-tion's objectives. Begin to identify risks through workshops or interviews with executive management and by focusing on strategies and related business objectives.
- Assess risks. Through risk assessments, try to reach a consensus on the impact and likelihood of each risk.
- Develop action plans and assign responsibilities. Every risk must have an owner somewhere in the organization.
- Maintain flexibility to respond to new or unanticipated risks. Put a business occupational safety plan in motion.
- Use metrics to monitor the effectiveness of the risk management process where possible.
- Communicate the risks identified as critical. Circulate risk information throughout the organization. It is not acceptable to identify important risks and never communicate them to the appropriate people.
- Embed ERM into the organizational culture. Integrate the knowledge of risks into organizational planning, budgets, and the performance manage-ment system.

Organization-wide occupational safety requires compliance to multiple laws and regulations that have been enacted to make the work environment a safer place.

At the organizational level, the integration of an occupational safety management system should influence policy and management arrangements, as well as stress the importance of having a safe environment. From this perspective, as mentioned in this chapter, occupational safety is a management responsibility that is organization-wide, and it should not be a mere task for departments and/or specialists.

EXERCISES

Expand on the following statement: a systems management approach to safety arises from a desire to cover for the failings of each component by combining components into a single, comprehensive strategy, the whole of which is greater than the sum of its parts.

REFERENCES

Deming, W. E. (1986). *Out of the Crisis.* Cambridge, MA: MIT Center for Advanced Engineering Study.

Occupational Safety and Health Administration (OSHA). (1996). OSH Act, OSHA standards, inspections, citations and penalties. Available May 30, 2013, at http://www.osha.gov/doc/outreachtraining/htmlfiles/introsha.html

Tennant, G. (2001). *Six Sigma: SPC and TQM in Manufacturing and Services.* Farnham, UK: Gower Publishing.

Walker, P. L., and Shenkir, W. G. (2008). Implementing enterprise risk management. *Journal of Accountancy.* Available May 30, 2013, at http://www.journalofaccountancy.com/issues/2008/mar/implementingenterpriseriskmanagement.htm

8 Occupational Safety Leadership

Leadership motivates people. People do the work. They design, develop, and implement work processes. Leadership motivates people to follow and achieve process goals. Management controls systems by measuring the processes to achieve an efficient and effective flow of resources for productivity. Like management, leadership can be learned. Almost anyone has the potential to be a leader. Training and assessments generally focus on the potential leader's capabilities by examining three core skills: the conceptual ability to create a vision and mission of the organization, technical skill in work processes, and human skill in relationship building.

Vision is a key to leadership. Leaders are passionate about their vision. They have a zealous focus on organizational and personal goals. Executive-level leaders focus on vision and moving the organization forward. They inspire followers by creating follower ownership in the vision, with shared ownership of the vision enabling followers to act. They are willing to challenge the status quo and innovate, empowering others to achieve the organizational goals. Technical skill deals with knowledge of a specific work or activity. First-line leaders generally have more skill and are focused on mission accomplishment. As the leader rises through the organization, less technical skill and more organizational vision are needed. Human skill means working with people at all levels. Leaders model the way by living the values they expect of all persons in the group.

Traits of leadership excellence are intelligence, self-confidence, determination, sociability, and integrity. Although these are sometimes difficult to measure, people can enhance these traits and skills. Intelligence traits include strong communication skills (verbal and written) and the ability to properly notice events and direct an effective response to them. Self-assurance and positive self-esteem mark the leader's self-confidence. Determination measures the leader's drive to accomplish the mission and tasks. Sociability implies the leader must possess good interpersonal skills that create cooperative relationships. Integrity is the most important characteristic of a leader, as leaders without integrity lead only until the followers lose their trust in them. Leaders with integrity inspire confidence in their followers. No one cares how much you know until they know how much you care.

Leaders must be genuine and real. Their actions must be congruent with their values. They build strong relationships with others. Social skills are essential for influence, the major characteristic of leadership. The ethics of leaders impact the ethical climate of their organizations by taking into account the interests of the organizational culture. Ethical theory provides a set of rules or principles that guide leaders in making decisions. Ethics are central to leadership because of influence. Ethical leaders respect the values and opinions of others. They listen to subordinates and are open to opposing viewpoints. Conduct and character are two domains of ethical theory. Ethical leaders are open, tell the truth, and hold themselves and their followers accountable. They treat all equally and assure that all treatment is fair.

Leaders influence. Influence may come from the position rather than the individual. Weak leaders rely on the authority of their position to compensate for their lack of personal leadership traits and skills. Some resort to fear to influence others in a negative sense. Highly effective leaders influence with personal power. Personal power can be developed by identifying and cultivating key characteristics that make up the ability to influence. Influence comes from integrity. Leaders must have integrity. You must be trustworthy all of the time. Virtue-based theories focus on the leader's character. Practicing good values leads to increased virtuousness, which leads to a more effective leader.

Charismatic leaders are confident and have a dominant personality. Their strong moral values and desire to influence set a strong role model and trust in the leader. Charismatic leaders express confidence in their followers and articulate their goals and vision in such a way that followers accept their ideology. More than just charisma is required for transformational leadership. As leadership is about influencing others, transformational leadership focuses on the morality of the leader and on directing activities in a positive way. Transformational leaders must have a clear vision of the future and communicate that vision in understandable language that is also energizing. They are the social engineers of the organization, shaping the culture of the group. They create trust through transparency: making their positions known to all, and concentrating on their strengths rather than dwelling on their weaknesses. Transformational leaders motivate followers to reach their full potential.

Transformational leadership emphasizes influence and inspirational motivation. The individual is connected to the goal of the leader through emotion and logic. Idealized influence describes the role-modeling process for followers. Strong moral and ethical standards cause followers to imitate them. Inspirational motivation occurs when the leader clearly communicates the vision and sets high expectations of the followers. Intellectual stimulation encourages creativity and innovation. Followers learn to challenge their own beliefs and to recognize personal biases that negatively impact mission goals. Individualized consideration focuses on the follower with the leader acting as coach, helping the follower achieve his or her full potential.

In servant, or follower-centered, leadership, leaders strive to serve others. They help others achieve their goals, thereby nurturing the follower with the vision of the leader. Their decisions consider their impact on the welfare of all. Servant leaders reach beyond their aspirations, placing the goals of the group above personal goals. The leader makes sure the followers have the resources necessary to accomplish their goals.

Within the various attributes mentioned in this chapter, leaders' skill is directed toward individual attributes, competencies, and leadership outcomes. Individual attributes of the leader are cognitive ability, motivation, and personality. Competency includes subject matter knowledge, problem solving, and social judgment. Leadership outcomes are observed in their ability for problem solving directed toward the forementioned performance criteria.

Career experiences and environmental factors influence individual skills. Career experiences include hands-on experience, the ability to handle challenging assignments, and the ability to mentor new employees and provide appropriate training for subordinates. Environmental influences are factors outside the person's control, such as outdated technology.

Situational leadership is focused on directive and supportive behaviors. Depending on the "situation," leaders can change their style for an ideal fit to the followers' needs. Such leaders assess the followers' competency and commitment to determine what style of direction is needed. The leader selects a style to match the followers' development level: high directive–low supportive, high directive–high supportive, high supportive–low directive, or low supportive–low directive behaviors. Development levels are also broken into four areas: low competence–high commitment, some competence–low commitment, moderate or high competence–varied commitment, and high competence–high commitment.

Take the example of a new employee, who may have low competence but high commitment. The leadership behavior most appropriate would be high directive–high supportive. The new employee needs direction to help him or her become more competent. As the employee develops competence, the leader moves to coaching, with the leader continuing to give direction and adding high support to the process. Eventually, the employee develops high competence and the leader can delegate the tasks totally to the employee.

Situational leadership has survived the test of time and proven to be an effective means of training leaders within an organization. The system works by having the leader apply the proper style to the individual follower. It is easy to understand and apply. A concern of the system is that people don't always fit nicely into one of the categories. With practice, the leader can learn quickly how to adapt between behaviors.

Contingency theory is similar to situational leadership in that leader success depends on the leader's ability to match his or her style with the situation. Contingency theory is supported with research that validates its ability to explain how effective leadership can be achieved. Like situational leadership, contingency theory is not tied to one "best" way of handling a situation, but rather teaches the leader to be flexible in his or her approach. Contingency theory sees leaders as either task motivated or relationship motivated. Three factors—leader–member relations, task structure, and position power—are key in using the contingency theory model. The emphasis of leader–member relations is how work teams get along. Task structure examines the specificity of instructions and task requirements. Position power is the authority of a leader to reward or punish followers (Northouse, 2006).

The path–goal theory looks at the leader's style, the characteristics of the subordinate, and the work situation. Path–goal assumes that subordinates are motivated when they are capable of performing the work, believe their efforts will result in a specific outcome, and feel the reward for the work is worthwhile. Path–goal looks at the components of leader behaviors, subordinate characteristics, and task characteristics to determine the leader's impact on subordinates' motivation. The theory works by the leader looking at the needs of the subordinates, and fitting her style to that need.

Leader behaviors are directive, supportive, participative, and achievement oriented. Directive leaders set clear standards of performance and make the process rules clear. Supportive leaders are approachable and show concern for the subordinates' needs. Participative leaders invite subordinates to share in decision making. Achievement-oriented leaders challenge subordinates to perform work at the highest levels possible. As with situational leadership, the leader in path–goal theory will adapt his or her behavior to the needs of the subordinate (Northouse, 2006).

Subordinate characteristics are a need for affiliation, a preference for structure, a desire for self-control, and task ability. Leaders' effectiveness depends on how they respond to subordinates with varying degrees of each characteristic.

Task characteristics provide motivation for subordinates through the task, formal authority, and the work group. The focus of the leader is to help subordinates through and around obstacles.

Leader–member exchange theory looks at leader–follower relations, being aimed at individuals rather than at a group as a whole. While the workgroup has many members, the leader's success is focused on one follower at a time. Each subordinate has different characteristics that require the leader to tailor an approach to the individual. In a workgroup, subordinates may be seen as being part of the "in" group or part of the "out" group. Personality plays a role in this process as a bond between the leader and follower grows or wanes.

Leader–member exchange theory focuses attention on building relationships between the leader and followers, and between followers. The theory shows the importance of communication in leadership. The "in" group works better together and is usually more efficient. Members of the "out" group are less effective. Leaders need to be aware of this and strive to nurture those who are "out."

The study of leadership examines work teams and explores ways to make teams more efficient for the task for which they were created. Sometimes, a work group is called a *team*, but a group with a lack of purpose and community is no more than a collection of individuals. Teams are individuals who band together and jointly strive to achieve a common goal. The study of team leadership supposes that individuals other than the formal leader can perform critical leadership functions.

The leader is basically a medium for processing information. The team comes together for the purpose of a unified effort in processing necessary information. Characteristics of team effectiveness include a unified commitment, a results-driven structure, a collaborative climate, and external support. The team *leadership model* integrates mediation and monitoring concepts to a group rather than to an individual.

Effective team performance begins with a mental model of the situation. The team leader develops a mental picture of the situation and relates it to the team. The team then takes action to solve the problem. The team leader must decide to intervene in the team's efforts by asking three questions:

1. *Should I monitor the team or take action?* The leader watches for internal or external factors that need attention, then must decide if his or her input is needed.
2. *Should I intervene to meet task or relational needs?* When the answer to question 1 is to take action, the leader focuses on the need to be addressed: is there a problem with the task, a problem with group relations, or both?

3. *Should I intervene internally or externally?* Internal support is needed at times when you must clarify goals, facilitate decision making, and emphasize standards of excellence. External support is when you must work in the organization to give the team credibility. External support may be gathering information from other functions so the team can function efficiently.

According to the psychodynamic approach, learned patterns of family dynamics influence leadership. Insight into one's own personality is thus beneficial. This approach is criticized because it is based on the psychology of the abnormal and focuses primarily on the personalities of leaders and followers. As such, it does not lend itself to traditional training.

Transactional analysis maintains that there are three ego states: parent, child, and adult. We shift in and out of the states in relationships. Transactions are complementary when they match the ego state of the other party (e.g., an adult ego state is matched by an adult ego). Freud identified three personality types as erotic (the desire to be loved), obsessive (requiring order and stability), and narcissistic (takes pride in personal accomplishments). A fourth personality type was added by Eric Fromm: marketer (adaptable to change). Carl Jung expressed four dimensions in assessing personality: extraversion versus introversion, sensing versus intuiting, thinking versus feeling, and judging versus perceiving. These can be combined into sixteen combinations. Leaders should learn their style; understanding your style will help you develop your skills to become more effective.

LEADERSHIP AND DECISION MAKING

Leaders are expected to make decisions that further organizational goals. Leaders need the right information to make effective decisions. The leader needs to know what is going on. The trust gained in leadership encourages others to provide the leader with up-to-date information, correct information, and sufficient information to make an informed decision about the work process. The leader can assess the employees to identify their organizational concerns. Leaders should identify processes that need improvement and gather data to support a change or the status quo. The data will show if an issue is truly a cause for concern and will help the leader prioritize multiple issues.

Some issues may be evident but misleading, such as groupthink. Groupthink is a problem where one or more people make a comment that suddenly everyone believes at face value without the support of data. Meetings to gather data and discuss options should be open to involved stakeholders, with all present able to voice concerns with the understanding that each concern should be supported with information before any change is made. Brainstorming is a process that allows people to come together in such an environment. Participants' statements are not challenged or supported, but accepted as a means of identifying the root cause of a problem. In fact, proper brainstorming techniques encourage others to build on previous comments. Another method similar to brainstorming is the Delphi technique, where stakeholders submit anonymous suggestions to a facilitator or bulletin board for the purpose of eliciting new ideas or solutions to a problem. It is used as a consensus builder by having multiple rounds of questions to get the parties to agree.

An opposing method for decision making is constructive conflict. This method puts opposing views at odds and forces them to prove their position and often disprove or refute the opposing position. The leader acts as judge, weighing the evidence from one side against the other, and asking questions to clarify points to come to a conclusion on the best course of action to take. Regardless of the method used, a decision must be made. All decisions should follow certain key values.

The decision must align with the mission and goals of the organization. Ask, "What are we really trying to accomplish?" It may be easy to make a decision with short-term consequences, but the leader must decide with the organizational mission and vision in mind. Leaders must also align their decisions with the core values of the organization. While a decision may further the mission, it may conflict with a stated core value. In the event that two core values are in opposition, leaders must prioritize the values to assure that the highest value takes precedence.

Next, leaders must evaluate the importance of a decision so they can allocate the appropriate resources and time to it. Leaders must consider who will be impacted by the decision, both inside and outside the organization. As leaders progress with a decision, they must weigh all options. There may be obvious solutions, but new information and technology may prove other solutions to be more viable. This is where the brainstorming or Delphi methods may help identify stakeholder concerns and buy-in for the decision from affected parties.

Now, make the decision. Not everyone, even with a consensus, will back you 100 percent. People fear the consequences of making a bad decision. If you make a bad decision, admit it, make the right decision, and move on.

EXERCISE

Expand on the following statement, and apply it to a security organization: manager and leader are two completely different roles, although we often use the terms interchangeably. Managers are facilitators of their team members' success, while a leader leads based on strengths, not titles. The best managers consistently allow different leaders to emerge and inspire their teammates to the next level.

REFERENCES

Northouse, P. (2006). *Leadership: Theory and Practice*. Thousand Oaks, CA: Sage Publications.

9 Comprehensive Risk Assessment for the Occupational Safety Manager

For occupational safety managers, risk assessment is an essential component of their vocation. A risk assessment is the primary tool to allocate resources in the protection of organizational assets. The risk assessment will look at organizational assets, accounting for both the criticality and vulnerability of the asset to determine the safety investment. The comprehensive process of risk assessment is a process of critical thinking. The risk assessment is a methodical process of evaluating credible threats, identifying vulnerabilities, and assessing potential consequences. Risk assessments will be the basis for protecting organizational assets through the application of strategies and countermeasures to reduce the threat of, vulnerability to, and/or consequences from a loss event.

OVERVIEW OF RISK ASSESSMENT

The risk assessment is a primary component of the risk management process. The objective of a risk assessment is to identify risks to organizational assets and propose an achievable level of protection that is commensurate with the level of risk, without exceeding that level of risk so that it will be cost effective. Risk is a function of the values of threat, vulnerability, and collateral damage via loss occurrence. The objective of risk management is to create a level of protection that mitigates vulnerabilities to threats and their potential consequences, thereby reducing risk to an acceptable level. Ideally, all risk would be eliminated. However, in practicality the elimination of risk is not feasible (Department of Homeland Security, 2008).

The safety manager, in consultation with organizational leadership, plays a critical role in safety decision-making processes. To make an informed risk-based decision regarding the mitigation or acceptance of risk, collaboration between the safety manager and organizational leadership is required. For any countermeasure that is recommended, the safety manager must provide all information pertinent to the decision, including the nature of the threat, specific vulnerabilities, an understanding of the potential loss consequences, and the costs.

The process of risk assessment will not prevent adverse events from occurring; however, it enables the safety manager to focus on those things that are likely to bring the greatest harm, and employ approaches that are likely to mitigate or prevent those incidents. Therefore, the risk assessment is not an end in and of itself, but rather is part of organizational practices that include planning, preparedness, program evaluation, process improvement, and budget priority development. The value of a risk assessment is not in the determination of a particular course of action, but rather in the ability to distinguish between various choices within the larger context.

The assessment of risk should not necessitate a comprehensive risk assessment for an entire organization. There is not a specific risk assessment methodology that transcends all organizational environments. There are going to be differing risk assessments for differing organizational environments. However, all risk assessments should be rooted in a critical thinking methodology. Risk assessments should employ a reflective reasoning process based on an understanding of the organization and its environment. The assessment chosen should adhere to the fundamental principles of a sound risk assessment methodology. The methodology should follow the premise of critical thinking in the following ways:

1. The methodology must be credible, and it must assess the threat, consequences, and vulnerability to a specific loss event.
2. The methodology must be reproducible, and it must produce similar or identical results when applied by various safety professionals.
3. The methodology must be defensible, and it must provide sufficient justification for deviation from the baseline.

A risk assessment is not going to be comprehensive to all organizational environments. Different environments require differing risk methodologies. In practice, various methodologies will provide varying outputs, from numbers and percentages to qualitative ratings. The safety manager must determine what outputs from the respective methodologies correlate with the desired level of protection. In organizations where multiple risk assessments may be conducted, the safety manager will need to evaluate the comprehensive findings and determine what countermeasure recommendations to implement from multiple risk assessments.

The risk assessment should also document the existing level of protection provided by the inherent qualities of risk mitigation, such as geographical isolation or existing safeguards. Levels of risk determined for each undesirable event should be considered in terms of inherent mitigation by existing countermeasures that provide a commensurate level of protection, meaning the higher the risk, the higher the level of protection.

The Department of Homeland Safety (DHS) defines a risk assessment as a product or process that collects information and assigns values to risks for the purpose of informing priorities, developing or comparing courses of action, and informing decision making, via the appraisal of the risks facing an entity, asset, system, network, geographic area, or other grouping. Put simply, a risk assessment can be the resulting product created through assessment of the component parts of organizational risk (U.S. Government Accountability Office, 2011b).

A risk assessment is a decision-making tool. It serves the safety manager by identifying and analyzing threats and vulnerabilities, and provides the basis for recommending appropriate countermeasures. The decision to implement those recommendations and mitigate the risk, or to accept risk, is that of safety management in consultation with organizational leadership. The safety manager, in consultation with organizational leadership, is responsible for identifying and implementing the countermeasure appropriate for mitigating a vulnerability, thereby reducing the risk to an acceptable level.

The risk assessment can be processed in a qualitative or quantitative framework. The qualitative risk assessment will be based on methods, principles, or rules for assessing risk based on nonnumerical categories or levels. An example would be where the safety manager assigns categorization of risks as low, medium, or high based on known data or assumptions regarding the risk environment.

A quantitative risk assessment will utilize a set of methods, principles, or rules for assessing risks based on the use of numbers where the meanings and proportionality of values are maintained inside and outside the context of the assessment. For example, a safety manager could use a quantitative risk assessment methodology to assess the risk of loss, based on statistical past loss data. While a semiquantitative methodology also involves the use of numbers, only a purely quantitative methodology uses numbers in a way that allows for the consistent use of values outside the context of the assessment. To make an informed risk-based decision regarding the mitigation or acceptance of risk, collaboration between the safety manager and organizational leadership is required. For any countermeasure that is recommended, the safety manager must provide all information pertinent to the decision, such as the nature of the threat, the specific vulnerabilities, an understanding of the potential consequences, and the costs.

ASSESSING RISK

The safety manager and organizational leadership need accurate information in order to make effective decisions. Safety management must have the authority, appropriate clearance, and access to expert resources to gain an understanding of risk to render sound decisions. This requires an understanding of the safety issues, and also of the mission and priorities of organizational leadership and the associated cost implications. This approach promotes comparability and a shared understanding of information and assessment in the decision process, and it facilitates better structured and informed decision making. The risk assessment is a methodical process, and it should do the following:

1. Define the organization and its context.
2. Identify risks to the organizational asset.
3. Determine asset vulnerability to identified risks.
4. Develop multiple approaches to deal with organizational risk.
5. Decide and implement the optimal approach to deal with asset risk.
6. Monitor and evaluate the risk mitigation measure(s).
7. Use risk communications. (U.S. Government Accountability Office, 2011a)

DEFINE THE ORGANIZATION AND ITS CONTEXT

To execute a risk assessment, it is critical to define the context that the risk management effort will support. The safety manager must gain an understanding of the environment in which the risks are to be managed, taking into account all organizational concerns and risk tolerance. Defining the context will inform and shape successive stages of the risk management cycle.

To determine the necessary level of protection to adequately mitigate risk, a safety manager must understand the role that assets play in an organization. Variations in the nature of mission, location, and physical configuration of an organization may create unique risks or risks that are relatively higher or lower in some cases than with other organizations. The baseline level of protection may not address those risks appropriately. It may provide too little protection, leaving an unmitigated risk, or it may provide more protection than is necessary, resulting in the expenditure of resources where they are not needed. This might reduce the availability of resources that could be applied elsewhere.

This initial step should clearly define goals and objectives that are essential to identifying, assessing, and managing areas that may impact the success of your risk assessment.

This concerns organizational leadership, the mission, and values of the organization. This step should also make considerations for existing risk management tools, including organizational policies and procedures. The safety manager must understand the decision-making process in the organization and the commitment of organizational leadership to support the risk assessment process.

Organizational leadership and staff must be invested in the risk assessment to assure their support and commitment to the process. The safety manager must also understand that organizational leadership must understand the technical aspect of the process so they will not only understand but also buy into the risk assessment process. This initial portion of the risk assessment process requires communication from the risk manager and all parties involved in the process.

In this initial step of risk assessment, the safety manager must identify the staff (including their skill levels), financial resources, and other organizational resources available for risk assessment efforts. The safety manager must be flexible and execute a great deal of understanding as he or she moves into the risk assessment process. The inability of a safety manager to recognize aspects of organizational culture and decision makers can have a negative impact on the success of the process.

The safety manager must also consider existing safety practices and the overall culture of safety in the organization. The safety manager should define in detail the following key areas of safety management:

- Current asset classification practices
- Existing assignment of roles for organizational ownership of assets
- Existing asset responsibilities, including tasks and procedures to be followed by the staff members who work with the asset
- Existing policies regarding the securing of organizational assets, including how violations are reported and dealt with
- Existing safety awareness practices, including staff educational and training programs related to asset safety

IDENTIFY RISKS

Identifying a preliminary list of risks can generally be done from a basic knowledge of the asset, its function in the organization, and an understanding of its vulnerabilities. This can be completed by understanding the asset, its function relative to the organization's goals and the objectives, then by determining the risks, hazards, resources, and organizational vulnerabilities that impact them, resulting in risk. This method will provide a list of potential loss situations to the organizational asset.

Risks to organizational assets come from both internal and external vulnerabilities. Examples of internal risks are often surrounding actions (or inaction) by staff and failures in organizational systems. All organizations are vulnerable to internal risks. This step of the risk assessment will identify weaknesses in the organization's internal systems and processes.

External sources of risk can be caused by many factors. External factors resulting in vulnerability can include global, political, and societal trends, as well as hazards from natural disasters, terrorism, cybercrime, pandemics, transnational crime, and human-caused accidents. It is the role of the safety manager to think critically and recognize all the potentials for loss.

DETERMINE VULNERABILITY TO IDENTIFIED RISKS

The purpose of this step is to utilize a methodological framework to objectively determine asset vulnerability to identified risks. This step involves the following:

- Determining a formal risk assessment methodology
- Gathering data
- Executing the risk assessment methodology
- Validating and verifying the data
- Analyzing the outputs

Determining a Formal Risk Assessment Methodology

In determining a formal risk assessment methodology, the safety manager should try to keep it as simple as possible while providing as comprehensive and accurate data as possible. The more simplistic any process is, the less likely it is to fail or be misinterpreted. Simplicity and practicality will have an inherent value when you are "selling" your safety program to organizational stakeholders by allowing you to better explain technical loss data to both laypeople and experts.

Risk assessment methodologies can be either qualitative or quantitative, but when well designed, both types of assessments can provide results for a valid understanding of the asset risk environment. Failure to choose a valid risk assessment for the assets can result in the process being flawed, if not completely unreliable.

Gathering Data

The safety manager should gather data to provide practical and valid data in the assessment. There are a number of potential sources for risk information. Some of the most commonly used sources for risk assessments include staff interviews, historical data, models, simulations, and subject matter experts.

Many pieces of data are not known precisely. When an organization has not experienced a particular negative event, it may be oblivious to that risk and loss potential. It is crucial that the safety manager follow critical thinking guidelines. The assumptions and uncertainty in the inputs should be considered in each step of the assessment's methodology to determine how they affect the outputs. Uncertainty in the outputs should then be communicated to the decision maker, as well as the assumptions that underpin the analysis. It is also useful to consider the impact of the uncertainty and how sensitive the assessment of risk is to particular pieces of uncertain data.

Executing the Risk Assessment Methodology

The risk assessment methodology that the safety manager chooses to use is the framework that guides how the data are processed to provide valid results for decision making. There are many different methodical frameworks that can be used to provide the safety manager with an understanding of the risk environment and subsequent methods of protection and control to enhance safety. In the process of implementing the risk assessment, the safety manager should critically evaluate the assessment tool to assure its validity in providing the information needed.

As the assessment process is implemented, the safety manager should begin to gain a justifiable understanding of the risk environment. It is within this process that the safety manager should also develop and evaluate alternative means of protection. While approaches for developing and evaluating alternatives are as diverse as the problem sets, considerations may include the following:

- Reviewing lessons learned from relevant past incidents
- Consulting subject matter experts, best practices, and governmental guidelines
- Brainstorming with stakeholders
- Organizing risk management actions
- Evaluating options for risk reduction and residual risk

Validating and Verifying the Data

The safety manager should continually validate and verify the data collected to assure its soundness. The importance of validating and verifying data is fundamental to the risk assessment process because it is the foundation for making decisions and protecting an asset, among a number of alternatives in an uncertain environment. The key moment in the execution of any risk management process is when a decision maker chooses among alternatives for managing risks, and makes the decision to implement the selected course of action. This can include making an affirmative decision to implement a new alternative, as well as the decision to maintain the status quo.

Analyzing the Outputs

In analyzing the outputs from the risk assessment, the risk manager should evaluate and monitor performance to determine whether the implemented risk management options can achieve the goals and objectives of the organization. In addition to assessing performance, organizations should guard against unintended adverse impacts, such as creating additional risk or failing to recognize changes in risk characteristics.

The evaluation phase is designed to bring a systematic, disciplined approach to assessing and improving the effectiveness of risk management program implementation. It is not just the implementation that needs to be evaluated and improved; it is the actual risk reduction measures themselves. Evaluation should be conducted in a way that is commensurate with both the level of risk and the scope of the mission.

In practice, the determination of risks to assets rarely occurs linearly. Instead, safety managers often move back and forth between the tasks, such as by refining a methodology after some data have been gathered.

DEVELOP MULTIPLE APPROACHES TO DEAL WITH ORGANIZATIONAL RISK

The safety manager should develop multiple approaches to deal with organizational risk. Analysis of the risk assessment process is often enhanced by developing various models and techniques from various loss scenarios. The utilization of differing approaches is based on assumptions regarding how potential risks materialize against organizational assets.

The risk assessment process should yield multiple process and protection measures to enable timely and relevant mitigation of risks by monitoring predictive indicators, escalating information on increased risk exposures, and making risk-informed decisions in an integrated manner. The safety manager should also analyze the impact of those risk mitigation measures while considering the costs and benefits of those alternatives.

DECIDE AND IMPLEMENT THE OPTIMAL APPROACH TO DEAL WITH ASSET RISK

Risk management entails making decisions about best options among a number of alternatives in an uncertain environment. The key moment in the execution of any risk management process is when the safety manager, in cooperation with organizational leadership, chooses among alternatives for managing risks and makes the decision to implement the selected countermeasure(s). This can include making an affirmative decision to implement a new alternative, as well as the decision to maintain existing safety measures.

Cost considerations can be a key factor in decisions to implement countermeasures. Cost–benefit analysis is a cost analysis methodology used to justify safety expenditure. The cost–benefit analysis is a comparative assessment of the benefits from your safety measure and the costs to perform it, in relation to the financial impact resulting from potential loss to the asset. The safety manager should consider costs for all phases of the safety project. Costs may be one-time capital costs or recurring costs. The determining criterion in a cost–benefit analysis is that the safety benefit exceeds the cost. The amount that the benefit exceeds the cost is based on the safety manager's determination in consultation with organizational leadership.

Monitor and Evaluate the Risk Mitigation Measure(s)

The safety manager should monitor implemented safety measures and compare their effects to help influence subsequent risk management alternatives and decisions. The risk assessment will identify possible risks and the likelihood of occurrences rated in terms of impact or severity and probability. This enables the safety manager to develop response strategies and allocate resources appropriately. Safety management then ensures that risk assessments become an ongoing process, in which objectives, risks, risk mitigation measures, and controls are regularly reevaluated.

The safety manager should evaluate and monitor performance safety measures to determine whether they are achieving their goals and objectives. In addition to assessing performance, the safety manager should monitor for unintended adverse impacts, such as creating additional risk or failing to recognize changes in risk characteristics as a result of implementation of the safety measure(s).

The evaluation phase is designed to bring a systematic, disciplined approach to assessing and improving the effectiveness of risk management implementation. It is not just the implementation that needs to be evaluated but also the actual risk reduction measures themselves. Evaluation should be conducted in a way that is commensurate with both the level of risk and the scope of the organizational mission.

Risk Communications

The foundation for each element of the risk management process is effective communications with organizational leadership, staff, other stakeholders, and customers. Consistent, two-way communication throughout the process helps ensure that all those involved share a common understanding of asset risk and the factors involved to manage it. Effective communication is an essential element in executing plans and countermeasures and in explaining risks and safety management decisions. Such external communications should occur throughout the safety management process and should be considered integral to effective risk management.

CONCLUSION

The primary role of the safety manager is to organize the countermeasures that are generally considered to mitigate the risk from a particular undesirable event. Safety managers are aided by risk assessments that help them understand organizational risk. A risk assessment identifies a generic set of undesirable events that may impact organizational assets, and risk management relates them to the applicable safety measures.

By determining the necessary level of protection according to a risk assessment, it is possible to ensure that the most cost-effective safety program is implemented without waste or lingering vulnerability. Incorporating a cost–benefit assessment will provide additional insight to the safety manager. The safety manager should determine whether the countermeasures adequately mitigate risk in an economically acceptable manner.

In all cases, the risk management process should document clearly the reason why the safety measure(s) are necessary. It is extremely important that the rationale for accepting risk be well documented, including alternate strategies that are considered or implemented, and opportunities in the future to implement the necessary level of protection, with those findings communicated to all stakeholders in the process.

EXERCISE

Risk assessment is a methodical process that promotes comparability and a shared understanding of information and assessment in the decision process; it facilitates better structured and informed decision making. What are the seven steps of the formal risk management process, and how would you apply them to a local construction site? Give examples for each of the steps.

REFERENCES

Department of Homeland Security. (2008). DHS risk lexicon. Available May 31, 2013, at http://www.dhs.gov/xlibrary/assets/dhs_risk_lexicon.pdf

U.S. Government Accountability Office. (2011a). Homeland Safety: Applying Risk Management Principles to Guide Federal Investments. Report to the Committee on Homeland Safety and Governmental Affairs, U.S. Senate. Available May 31, 2013, at http://www.gao.gov/new.items/d07386t.pdf

U.S. Government Accountability Office. (2011b). National Preparedness: DHS and HHS Can Further Strengthen Coordination for Chemical, Biological, Radiological, and Nuclear Risk Assessments. Report to the Committee on Homeland Safety and Governmental Affairs, U.S. Senate. Available May 31, 2013, at http://www.gao.gov/new.items/d11606.pdf.

10 Computer and Information Security for Occupational Safety

Many organizations collect large amounts of information. The information can range from employee records to operational data, customer databases, trade secrets, and so on. Organizational information that is commonly collected may contain administrative reports documenting various types of data for day-to-day operations, or other supplementary support. The fact is that most organizations retain a lot of information in their operational processes. This chapter will use educational environments as examples for computer and information security; however, the concepts and examples are applicable to many organizational environments.

Information is a vital resource for organizational staff to utilize in planning programs and services, and in scheduling and completing reports for organizational stakeholders, if not local, state, and federal agencies as well. In emergency situations, information must be readily available to make a valid response to a life safety event. Organizations must maintain information integrity, accessibility, and confidentiality.

Organizational information is a compilation of records, files, documents, and other materials that may contain many sensitive pieces of data. Information may be kept in a variety of formats, including handwritten, printed, and digital files and video and audio recordings. Information must be accurate and available to make timely decisions. In emergency situations, accurate and accessible information may play an important factor in safety. Furthermore, organizations could have legal consequences if they fail to safely maintain information.

This chapter has three primary purposes. First, it is to serve as a guide to help organizations look at information and give it a value. A risk assessment is a methodological tool to help practitioners look at information as assets to the organization and determine a value of that artifact, then understand the threats and vulnerability to it as the basis for its protection. Second, this chapter will address legislative requirements outlining methods for keeping information safe. Finally, fundamental principles in applying policy and procedures in the pursuit of safekeeping of information will be addressed.

SAFEGUARDING INFORMATION

Protecting information requires maintaining confidentiality and integrity, while ensuring accessibility and availability of information via security, safety, and ethical behavior. Confidentiality involves the prevention of disclosure of information to unauthorized individuals. Information integrity means that data cannot be modified without someone being able to detect its modification. Information availability means that information must be accessible when it is needed. The information is made safe by providing physical and procedural measures, ensuring that only those with specific needs have access to the information, and promoting an ethical understanding to all involved as to the sensitivity of the information.

Organizations should develop plans to safeguard information. Developing a comprehensive information safety plan requires going through a process to assess information value, vulnerability, and risk. This process is called a risk assessment or risk analysis. The risk assessment is a continual and dynamic process as risk management is an ongoing iterative process. Risk assessment must be repeated indefinitely. Anytime that new technology is introduced or changes are made to informational processes, a school's information environment changes. These changes present the possibility of new threats and vulnerabilities impacting the safekeeping of information. Therefore, the risk must be reevaluated. The organization must constantly balance countermeasures and controls to manage risks to information. Information safety involves balancing between productivity, the cost-effectiveness of the countermeasure, and the confidentiality, integrity, and availability of the information.

An information risk assessment is a variation of a more traditional risk assessment, which focuses on identifying assets and looking at the threats and vulnerabilities of each asset to potential loss. In conducting an information risk assessment, the first step is to identify information assets and estimate their value to the organization.

Identifying information assets requires investigation and asking questions such as "What type of information does the organization maintain?" "What is its value to the organization?" and "Where is the information stored, whether digitally or in hard copies?"

Information includes all the data maintained by organizations and by other parties acting for the organization. Information collected by an organization about staff often includes personal information, such as pay data, Social Security numbers, pictures, or a list of personal data that can have varying degrees of sensitivity (e.g., records pertaining to medical and health information of staff), and these should also be addressed as sensitive information. The organization will also maintain information that may be of a more sensitive nature. This could include information specifically related to organizational processes.

The security manager should identify information and place a value on it. The security manager should calculate the value of each information asset. This would require a qualitative or quantitative analysis of information to determine what that information is worth to the organization. Questions such as "How can we meet the primary mission of this organizational enterprise in the absence of this information?" and "What legal ramifications exist if we do not safely maintain this information artifact?" should be considered.

The second step is to conduct a threat assessment to the organization's information. A threat assessment of information is a daunting task for educators. It requires the knowledge of all the information maintained by the organization and potential threats to that information. Threats can be obvious, such as natural disasters, or more ambiguous, such as a computer virus. Information stored in different media may have completely different characteristics and subsequent threats. The threats to the file cabinet in a secretary's office are very different from those to organizational information stored on the network cloud.

A threat assessment should look at each information asset and then determine its liability. Threats can include both accidental and malicious acts originating from inside or outside the organization. The threat assessment dictates that each situation of concern be viewed and assessed individually. Application of the assessment is guided by the facts of the specific threat and carried out through an analysis of its characteristics. Providing a safe information environment is very different and is based on the medium in which the information is stored.

Threats to digital information are viruses, malware, exploited vulnerabilities, remote access, mobile devices, social networks, and cloud computing. Viruses and malware can render information useless. Most active computer users have fallen prey to viruses and malware that damage and destroy information. Viruses are malicious codes that cause an infected computer to spread the virus to other associated computers via the network or email contact lists. Malware, which may be destructive or disruptive in nature, will not have the inherent ability to spread itself. So while you may get a malware infection by visiting a website that hosts the malware, it is not actually a virus unless it utilizes your contact list or network directory services to propagate itself. Either the malware- or virus-infected machine or network can make an organization's information useless (National Center for Education Statistics, n.d.).

A vulnerability exploit is another threat to an organization's digital information. A vulnerability exploit is where a "hole" is found in applications or systems software that facilitates unauthorized access to information. It will allow an unauthorized user to access information and data that they do not have legitimate rights to access. Vulnerability exploits are often utilized by viruses and malware to penetrate networked computer systems.

Mobile devices are another threat for information safety. Storing and removing sensitive information from an organization on a USB drive and then losing the drive poses a threat. Another good example is the laptop computer. Organization-owned laptops often house sensitive information. The loss of that information to theft is a major cause of the loss of information integrity. When computer information is outside the organization's physical network, its threats change. Having an organization-owned laptop, with organizational information on it, outside the logical and physical security of the business poses additional threats from both loss and exploitation and must be considered in a threat assessment.

Online social networking poses additional threats to information safety. Social networking sites, such as Facebook and Twitter, can pose serious threats to organizational information both directly and indirectly. Social networking sites are breeding grounds for viruses and malware plus other attacks to organizational information systems. Social networking sites can also give organizational staff and stakeholders the ability to post sensitive information.

Educational organizations using cloud-based resources must be aware of threats caused by utilizing that resource. The cloud is an Internet-based network that allows students, faculty, and staff to access applications on a network platform that is leased or rented to the organization in a way similar to a utility or tenement arrangement. The cloud extends the system so that staff and stakeholders have 24/7 access to not only their email but also personal files and application software anywhere they have a computer and Internet access. For example, digitized information does not have to be installed on the remote computer, nor does it have to be connected to a local area network to function. The application is run from the cloud, and the data generated are stored on the cloud, which can be accessed anytime by anyone with Internet service and proper access credentials.

Staff may unwittingly pose a major threat to organizational information. This could range from a desperate and disgruntled employee who was just fired because of performance issues to the careless employee who unknowingly releases sensitive information. There is no way to completely eliminate the threat of legitimate insiders, but through good safety policies and procedures, information loss could be greatly minimized. Careless and untrained employees could unknowingly release sensitive, if not damaging, information about an organization. Policies, procedures, training, and technical measures can play a major role in reducing an organization's information threats (McCallister et al., 2010).

Threat assessment of information is a huge task. It requires the knowledge and skills to address threats to the information that is maintained by organizations. Regardless of whether a threat is from natural disasters or a computer virus, it must be addressed. And once the threat is addressed, the organization should evaluate the vulnerability of information to that loss.

The third step is to conduct a vulnerability assessment to organizational information. A vulnerability assessment should calculate the probability that a threat to information assets could occur. The vulnerability assessment could be based on quantitative or qualitative evaluation of threats to information. The resulting data should give you an idea of the likelihood that loss of confidentiality, integrity, or access may occur so that measures of information safety could be implemented (National Center for Education Statistics, n.d.).

Often the creation of a chart or matrix such as the one shown here will facilitate professionals in determining threats and vulnerabilities to information. Either qualitatively or quantitatively, educators can value information and the threat vulnerability (Elky, 2006).

High Threat Vulnerability			High Risk
Medium Threat Vulnerability		Medium Risk	
Low Threat Vulnerability	Low Risk		
	Low Information Value	Medium Information Value	High Information Value

For example, for each informational asset, give it a value from low to high. Then, for each threat vulnerability to the information, give it a value from low to high. The intercept of these characteristics should provide the data to make a decision regarding how the informational asset should be protected from that threat.

An overly simplistic example is the threat of electrical surges to render information stored on a computer irretrievable. In some parts of the country, this could be a very high threat vulnerability, and the information could be very highly valued. In these situations, it would be worth spending $20 for a surge protector. But is the value justified in buying a $500 surge protection system? From the data in this scenario, we do not know if we can justify spending more money because the data will have to be balanced in terms of a comprehensive information safety program and organization budgets. The point is that with the risk assessment, we can make value judgments based on more than assumptions.

The data from an accurate risk assessment of information value, threats, and vulnerabilities will provide the security manager with the ability to identify, select, and implement appropriate security measures. This does not require the security manager to be an expert in safeguarding information, but it does assume that he or she has knowledge of information and knows when to provide a proportional response, even if that requires getting expert assistance in completing the task. The security manager should understand productivity, cost-effectiveness, the value of the informational asset, its threats and vulnerabilities, and subsequently how to go about protecting it (Elky, 2006).

Implementing a plan to keep information safe in organizations is a balancing act. Security managers must evaluate the effectiveness of measures that ensure security without discernible loss of productivity. Providing information security is contingent upon the security manager understanding the environment, risks, and vulnerabilities of technology. Because of the breadth of the issues and complexity of the environment, there is no more valid approach to achieving an information-safe environment in organizations than to promote secure, safe, and ethically sound informational practices by all staff and stakeholders. An organization must have both policies and procedures to provide for authentication, firewalls, and virus protection on its computer systems.

Authentication will provide secure login and allow individuals to access only the data and applications they require. The firewall also serves as an access control device that will limit user access to unwanted programs and data. Virus protection programs will protect users from viruses and malware on the network. Each of these three areas must be addressed by computing professionals in that environment. The best role for the security manager is to be aware of the value of information and its associated risks to develop policies and procedures to protect it. For the security manager, this means working with a service provider or consultant because they usually have access to greater levels of expertise (Elky, 2006).

Information security in organizations is very necessary because of the liabilities that result from improperly handling information. In many organizations, federal regulations require that organizations must make information safe regarding who will interact with it and how. Organizations must understand areas of compliance and how to safeguard data to the standards and regulation requirements.

FAMILY EDUCATION RIGHTS AND PRIVACY ACT OF 1974 (FERPA)

Schools, for example, have comprehensive information security measures mandated by the Family Education Rights and Privacy Act of 1974, commonly known as

FERPA. FERPA is a federal law that protects the privacy of student education information. Students have specific, protected rights regarding the release of information, and FERPA requires that institutions adhere strictly to its guidelines. Therefore, it is imperative that the faculty and staff have a working knowledge of FERPA guidelines before accessing organization information.

With FERPA, educational information is put into two broad categories: directory information and nondirectory information. Each category of educational record is afforded different safety protections. Therefore, it is important for faculty and staff to know the type of educational record that is being considered for disclosure (Office of Family Policy Compliance, 2008).

Directory information is the student's educational record. Directory information is information that is generally not considered harmful or an invasion of privacy if released. Under FERPA, the organization may disclose this type of information without the written consent of the student. However, the student (or guardian until a student is eighteen years of age or attends a postsecondary institution) can restrict the release of directory information by submitting a formal request to the school. Directory information includes the following (Office of Family Policy Compliance, 2008):

- Name
- Address
- Phone number and email address
- Dates of attendance
- Degrees awarded
- Enrollment status
- Major field of study

Educational organizations should always make students and parents aware that such information is considered by the organization to be directory information and, as such, may be disclosed to a third party. Organizations should overstate to students and parents that they can prevent the release of directory information (McCallister et al., 2010).

Nondirectory information is any educational record not considered directory information. Nondirectory information must not be released to anyone without the consent of the student, or a parent or guardian until a student is eighteen years of age or attends a postsecondary institution and then at the consent of the student. Therefore, a student in post-secondary education environments regardless of age must consent to allow parents to access that data. FERPA insists that faculty and staff can access nondirectory information only if they have a legitimate academic need to do so. Nondirectory information includes the following (Federal Register, 2001):

- Social Security numbers
- Student identification number
- Race, ethnicity, and/or nationality
- Gender
- Transcripts
- Grades

Student's distinct identifiers, such as race, gender, ethnicity, grades, and transcripts, are nondirectory information and therefore are protected educational records under FERPA. Students have a right to privacy regarding information held by the school. Organizations must ensure that students have privacy with this information under FERPA. FERPA also gives students and parents (again, the latter if the student is under eighteen) the right to access educational information kept by the school. It also gives the right to limit the educational information that is disclosed, the right to amend educational information, and the right to file complaints against the organization for disclosing educational information in violation of FERPA (Federal Register, 2001).

Students and parents have a right to know about the purpose, content, and location of information kept as a part of their educational records. They also have a right to expect that information in their educational records will be kept confidential unless they give permission to the organization to disclose such information. Therefore, it is important to understand how educational information is defined under FERPA (Office of Family Policy Compliance, 2008).

Prior written consent is always required before institutions can legitimately disclose nondirectory student information. In many cases, organizations tailor a consent form to meet their unique educational needs. However, prior written consent must include the following elements (Federal Register, 2001):

- Specify the information to be disclosed
- State the purpose of the disclosure
- Identify the party or class of parties to whom the disclosure is to be made
- The date
- The signature of the student (or the parent or guardian of the student) whose record is to be disclosed
- The signature of the custodian of the educational record

Prior written consent is not required when disclosure is made directly to the student or to other organization officials within the same institution where there is a legitimate educational interest. A legitimate educational interest may include enrollment or transfer matters, financial aid issues, or information requested by regional accrediting organizations (Office of Family Policy Compliance, 2008).

Institutions do not need prior written consent to disclose nondirectory information where the health and safety of the student are at issue, when complying with a judicial order or subpoena, or where, as a result of a violent crime, a disciplinary hearing was conducted by the school, a final decision was recorded, and the victim has sought disclosure. In order for institutions to be able to disseminate nondirectory information in these instances, FERPA requires that institutions annually publish the policies and procedures that the institutions will follow in order to meet FERPA guidelines (McCallister et al., 2010).

The Family Education and Privacy Act was enacted by Congress to protect the privacy of student educational information. This privacy right is vested in the student. Institutions may not disclose directory information in the student's educational record without the student's consent. Institutions must have written permission from the student in order to release any information from a student's educational record (McCallister et al., 2010).

OTHER REGULATORY ACTS REGARDING INFORMATION

Information is protected by other agencies and groups in addition to FERPA. The Departments of Agriculture, Health and Human Services, and Justice defend information privacy and apply it to the information in organizations. State and local entities may require safeguards for handling information in a safe and secure manner. The professional standards of ethical practice in organizations representing professionals such as doctors and nurses, psychologists, and others may also establish privacy restrictions (McCallister et al., 2010).

For example, information regarding drug and alcohol prevention and treatment services are covered by confidentiality restrictions administered by the U.S. Department of Health and Human Services. Some states have regulations regarding employee rights to seek treatment for certain health and mental health conditions, including sexually transmitted diseases, HIV testing and treatment, pregnancy, and mental health counseling. Some state regulations also protect information pertaining to HIV confidentiality, medical information, child abuse, privileged communications, and state-specific information retention and destruction regulations (Privacy Rights Clearinghouse, 2010).

MAKING INFORMATION SAFE

Information can be a compilation of records, files, documents, and other materials that contain information maintained by organizations. Likewise, information is a vital resource for organizations in planning, in operations, and in providing services. Organizations can be legally and ethically challenged to maintain information integrity, accessibility, and confidentiality. Information may be kept in a variety of formats, including handwritten, printed, and digital files and video or audio recordings. Organizations have an obligation to maintain information integrity, accessibility, and confidentiality. Information must be accurate and available to make timely decisions. An organization's obligation could have legal consequence if it fails to safely maintain private information of students (Privacy Rights Clearinghouse, 2010).

Organizations have an ethical responsibility to keep personal information about its students safe. Public organizations are further bound by federal legislation that provides mandates regarding information. Therefore, it is imperative that organizations have a plan regarding information security.

The final part of this chapter addresses the role of the organization in providing a secure information environment. Providing information security is contingent upon security managers understanding the environment, risks, and vulnerabilities of technology. This chapter also assumes that most organizational information is digital. Because of the breadth of the issues and complexity of the environment, there is no more valid approach to achieving a safe information environment in organizations than to use a multimode approach. This approach will promote secure, safe, and ethically sound practices. Organizations can create information security by providing security, safety, and awareness of anyone who deals with that information. The most common approach is to use layers to protect information. There are specific tools that can be utilized at each of these layers.

Instituting security will require specific technical information that most organizational stakeholders will not possess. Because of this, most organizations will need an expert to install and configure these applications and appliances. Security layers would include firewalls and routers. A firewall will stop unwanted applications and data from entering an organization's network. Unwanted applications and data could be anything from known viruses to social networking sites. The firewall can block all network traffic that is unwanted. For example, if you do not want Facebook to be accessed from organizational computing resources, then add it to the blocked list.

The routers must be protected on the network. Routers are like a train station through which all information passes. They direct information where it needs to go. The router is like a firewall because it can stop information from being dispersed into the network; however, its primary function is to direct network traffic both in and out of the local network. From this perspective, a firewall or router can stop unwanted applications and data, but only the routers and switches show data where they need to go on the network. The router is a very critical link for a network.

Network controls are a third primary layer to address information safety. Network controls will include authentication and file-sharing controls. Authentication will provide a user name and password authentication system. This provides that only authorized users access the network and its information. The file-sharing controls set an access level for each authenticated user to access applications and information on the network. From this perspective, you give information users only the rights necessary to do their task. In regard to information, you can give user rights to read, write, and modify discriminately for each information item stored on the network. The keys for network controls are to have a good password system and give users rights to only what they need.

Software layers of safety will include virus and malware detection or prevention systems. Malware and viruses can be some of the biggest problems for digitally stored information. The best way to prevent losses from these threats is by maintaining adequate antivirus and malware protection software and keeping all updates installed on organization computers. Antivirus and malware software must be updated very frequently. Often, a new virus cannot be stopped until the antivirus software has a definition of the virus to stop or quarantine it. Oftentimes, an update to a computer's operating system and application software is required because the manufacturer has found vulnerability in the system that must be addressed through updates and patches.

The final layer of information security involves creating informed information custodians. Information users must understand the value of information to the organization. Through training and continual reinforcement, information must be respected and cared for to assure its integrity, accessibility, and confidentiality. Everyone using organizational information must adhere to an acceptable use policy and be aware of consequences for violating it. Instill a sense of information responsibility. The best time to begin establishing a culture of information security is now.

It is was not the intent of this section to address information security from the view of a network administrator, but to make the security manager aware of what is needed to create and maintain an information-safe environment in the organization. Organizational stakeholders should understand a fundamental process such as an information risk assessment to view the value of, threats to, and vulnerability of information. An organization must be aware of all federal, state, and local laws regarding the use and maintenance of information. And, finally, organizations must have both policies and procedures to institute safety systems for authentication, firewalls, and virus protection of computer systems. Only through these measures can organizations adequately address information security.

EXERCISE

Sensitive personal information of 105 Bolder Machine Works employees, including names and Social Security numbers, were accidentally emailed to several customers Monday. Bolders Machine Works director's office released the information of the employees, via an email attachment sent to 26 recipients. Plant Manager Mitch Johnson informed the employees whose information was disclosed. "We take the safety of employee information very seriously, and we deeply regret this error," Manager Johnson said. "We are reviewing the incident and are taking steps to prevent similar incidents from occurring in the future. This shouldn't have happened, and we apologize for it."

On further investigation the manager determined the information was sent to a group of customers as an attachment, when the intended file attachment was accidentally replaced by the employee data file by an administrative assistant. It was an honest mistake.

- As a safety manager how would you address this issue?
- What factors must be considered in the addressing the issue?
- What can be done do to make sure this does not happen again?

REFERENCES

Elky, S. (2006). An introduction to information system risk management. Available November 15, 2011, at http://www.sans.org/reading_room/whitepapers/auditing/introduction-information-system-risk-management_1204

Federal Register. (2001). 34 CFR Part 99, Part V, Family Education Rights and Privacy, Final Rule. July 26.

McCallister, E., Grance, T., and Scarfone, K. (2010). Guide to protecting the confidentiality of personally identifiable information (PII). NIST Special Publication 800-122. Available November 15, 2011, at http://csrc.nist.gov/publications/nistpubs/800-122/sp800-122.pdf

National Center for Education Statistics. (n.d.). Data security checklist. Available November 15, 2011, at http://nces.ed.gov/programs/ptac/pdf/ptac-data-security-checklist.pdf

Office of Family Policy Compliance, Family Education Rights and Privacy Act (FERPA). (2008). FERPA primer: the basics and beyond. Available November 15, 2011, at http://www.naceweb.org/public/ferpa0808.htm

Privacy Rights Clearinghouse. (2010). Fact Sheet 29: Privacy in education: guide for parents and adult-age students. Available November 15, 2011, at https://www.privacyrights.org/fs/fs29-education.htm#11

11 Cybersecurity

Cybersecurity and cybersafety are topics that have spawned from the computer security domain since the profusion of Internet-based systems. Businesses rely on a vast array of computer networks to communicate, plan, provide services, and literally run our economy. The "cyber" aspect denotes the ability to remotely, if not virtually, access an organization's networked computer systems.

Cyber intrusions and attacks on networked computer systems have increased dramatically over the last decade, exposing sensitive personal and business information, disrupting critical operations, and imposing high costs on the economy (Department of Homeland Security, n.d.).

There is an increased vulnerability to network systems that have been deemed as critical infrastructures such as financial systems, chemical plants, and water and electric utilities. The role of security management in dealing with these environments often becomes the shared responsibility of various organizational units. Cybersecurity takes a multidimensional approach because most organizations have administrative technology systems to retain everything from employee data to inventories to process data used in the daily operation of the business. And, challengingly, a security manager will have a role in facilitating a secure cyberenvironment because, logistically, cybersecurity extends from the organization's local area network to the far-reaching network of the Internet.

Computers and cybertechnologies enhance organizational performance by engaging, involving, and empowering stakeholders. Further diffusion of Web 2.0 technologies and smart phones provides another interesting dimension to cybersecurity. Staff as well as organizational leadership have the ability to access, communicate, and process data via a robust network platform that they can hold in a hand or store in their pocket. Web 2.0 has made the Internet an instantaneous, participatory, and interactive community that differs from the initial incarnation of the Web as an environment where users could gather (or post) information and communicate via asynchronous systems. Smart phones, utilizing mobile computer operating systems, provide users with the ability to interact via Web 2.0 applications synchronously. Web 2.0 combined with smart phones have provided advanced computing and communication tools to their users and an additional challenge in the maintenance of cybersecure environments.

The role of organizations is increasingly challenged to maintain a technological environment that is cybersecure to all stakeholders, including staff and leadership. In addressing the assurance of a cybersecure environment, the first section of this chapter will be viewed from three cyberenvironments: the Internet, social networking, and smart phones, each with their own purpose, attributes, significance, value, and subsequent risks. The second section focuses on creating a cybersecure environment by implementing measures to cultivate secure and ethical cyberpractices.

Today's developing information age technology has intensified the importance of cybersecurity, which is now as critical as physical security in protecting many organizations. The energy sector has rapidly responded to the increasing need for enterprise-level cybersecurity efforts and business continuity plans. Vulnerability assessments have not only improved security but also demonstrated organizational commitment to secure organizational environments through cybersecurity.

It is not the intent of this chapter to provide the ability to develop a cybersecure environment in organizations. Technical aspects such as firewalls, malicious code, and access controls need to be addressed by network administrators who can configure those protection systems properly. However, it is proposed that this chapter will enlighten security managers with a knowledge base that will allow them to understand the various factors that construct an organization's cyberenvironment, so that they can provide an understanding to articulate an approach that can reduce an organization's cyberrisks. Minimizing cyberrisks in an organization can be best addressed by making staff and stakeholders aware of their role in the process.

COMMON CYBERENVIRONMENTS

THE INTERNET

The Internet is a worldwide public computer network. The Internet was originally founded by the U.S. Department of Defense in the 1970s using Transmission Control Protocol/Internet Protocol (TCP/IP) to connect computers and networks. Since the inception of Hyper Text Markup Language (HTML) in the early 1990s, the Internet evolved from a text-based communications platform to graphically interfaced Web pages. HTML fueled the development of the World Wide Web (WWW) into websites that would be publicly accessible, would be hosted via Internet-connected network servers, and would allow for other media files rather than mere text to be accessed.

The use of the Internet in organizations grew rapidly after the inception of HTML. The Internet has become a paradox to the organization. The Internet is a great resource, full of endless amounts of information and resources. Staff use of email has diffused throughout all organizations in the last twenty years. Staff Internet access is not only in the organization setting; research shows 82% of Americans have broadband Internet access at home (Rainie, 2011). While the Internet provides many positive attributes for organizations, it is easy to find sites that could have a negative impact. The paradox is that while the Internet is an environment that is beneficial to modern business systems, it also contains many risks and hazards to organizations.

EMAIL

Email in an organization is frequently used by staff and other stakeholders for sending and receiving electronic messages. Email allows employees to keep in touch with family, friends, and peers, as well as engage in business-related communication.

Potential risks with email relate to the inherent quality of the Internet. It utilizes a public network architecture where you can communicate with other email users indiscriminately. There is no required validation that users are who they claim to be. Furthermore, email users can communicate with others through unsolicited messages. These unsolicited messages, or spam, are varied in content and often promote sexually explicit material, products for sale, or money-making schemes. They can also act as a host for a malicious program.

Both Microsoft and Google are providing organizations with free email as well as online communications, applications, and storage. Many organizations use the services of cloud-based networks because of their low cost and the limited personnel needed to provide this enhanced service to the organization. There are, however, risks associated with movement to cloud-based email systems. While it is easy to monitor servers when they are run by internal data centers and under the control of the organization's IT department, it is more difficult when you have little control over servers that are located somewhere in the cloud. Therefore, it is important to measure and analyze not only performance but also the security of hosted information. It is very important that organizational leadership understands this and knows how to respond to risks.

INTERNET BROWSING

Internet browsing provides the means to explore information on worldwide computer networks, usually by using a browser such as Microsoft Internet Explorer, Google Chrome, or Firefox. The browser allows access to rich educational and cultural resources (text, sounds, pictures, and video). This also gives users an improved ability to understand and evaluate information and stay informed by accessing websites.

Risks associated with Internet browsing relate to sites with inaccurate, misleading, and untrue information. There is also access to sexually explicit images and other sites promoting hatred, bigotry, violence, drugs, cults, and other things not appropriate in the organizational setting. The Internet in general has no restrictions on marketing. Some Internet sites deceptively collect personal information in order to sell products via requests for personal information for contests and surveys. The Internet is a relatively wide-open interface to share data without any form of censorship.

ONLINE CHATTING

Online chatting is a popular communications tool used by many. Online chatting is reading messages from others as they are typing them, usually in a theme-specific or social network–specific interface. Chatting's inclusion in social networking sites has helped to maintain its appeal as a widely used communications medium. Chatting is popular because it allows users to communicate with people from around the world by synchronous typing of text into a chat interface. Staff can connect to others via websites or social networking portals, which in itself provides inherent risks.

Chatting is a risky environment because it provides an interface where people can communicate, in real time, with as much anonymity as they desire. Social networking sites add a new dimension to chatting because of the ability to have an online profile. Online profiles make searching differing demographic groups simple and easy. Hackers will befriend users via social networking sites and gain their trust by behaving as an understanding and trusted friend. Once trust is gained in a chat environment, staff can be susceptible to illicit activities.

SOCIAL MEDIA

The rise of Web 2.0, and most notably online social media, has had a profound effect on society and organizations as well. Web 2.0 combined with broadband Internet access has changed the way people communicate, process, and store data, and the diffusion of Web 2.0 has impacted users across all demographic groups. Web 2.0 technologies include social networking sites, blogs, wikis, video-sharing sites, hosted services, Web applications, and tags. For users, these tools, which are typically free or low cost, represent a transition from institutionally provided to freely available technology.

Web 2.0 technologies are possible because of the adaption of programming tools, such as asynchronous JavaScript (AJAX). AJAX is a group of interrelated Web development tools that are used to program interactive Web applications. This programming aspect, combined with diffusion of residential broadband Internet access, created an environment that evolved and migrated users to a very interactive form of the World Wide Web that facilitated interactive social networking. This AJAX-infused adaptation of the World Wide Web is now simply referred to as Web 2.0.

Web 2.0 is the current rendition of the World Wide Web that provided a "social" approach to generating and distributing Web content, characterized by open communication, decentralization of authority, and freedom to share and reuse information (Acar, 2008; Madge et al., 2009; Subrahmanyam et al., 2008). Socializing through Web 2.0 interactions is exemplified by photo and video sharing (e.g., Photobucket.com, Flickr.com, and YouTube.com), wikis and blogs, (e.g., Wikipedia.com, TripAdvisor.com, and UrbanSpoon.com), as well social networking sites (e.g., Facebook and Twitter).

Online networking has proliferated throughout the Web 2.0 environment because of an interactive design attribute that has been further propagated by increased residential Internet access and bandwidth (Acar, 2008; Madge et al., 2009; Subrahmanyam et al., 2008). The World Wide Web today has developed into a network of participation that typifies online social networking. There are a variety of online social networks in existence, the most popular being Facebook and Twitter.

Social networking sites combine common characteristics that allow profile creation, friends' listings, and public viewing of friend lists. Online social networking websites also allow users to create a unique Web presence, referred to as their *profile*. Through their profile, social network users live out an online identity while exploring friendships and relationships with other individuals who also have profiles on that website. Online social networking does not function entirely in real time, like conventional chat rooms and instant messaging, so the interactions that take place are not always instantaneous, even though most have chat room functionality.

Most social networking profiles are developed from responses to questions that request a user to disclose a variety of personal information (Mazer et al., 2007; Mitrano, 2008; Steinfield et al., 2008). Personal information includes user names or other identifiers such as sexual orientation, organizations, geographical location, and the extent of the relationship that staff members are currently in or are seeking with others. The profile also allows for self-expression through personal photographs and videos. Staff can also make available their list of friends an member groups, as well as create an area where individuals can post remarks or statements from others. An individual's social networking profiles have distinct web addresses that can be bookmarked or linked, allowing others to use and share that data with third parties.

The most popular online social network for staff is Facebook, an online directory that connects people through social networks. The website, www.facebook.com, was initially designed and developed in 2004 by Mark Zuckerberg, a Harvard University sophomore, and was inspired by a widely known paper version of a college "face book." The directory consisted of individuals' photographs and names, and it was distributed at the start of the academic year by university administrations with the intention of helping staff get to know each other.

Zuckerberg's initial intention was to create an online website to help Harvard coeds get to know one another for the purpose of finding roommates (Shier, 2005). It is no surprise that the Facebook website has grown in popularity among staff because it was initially designed exclusively for them.

From its creation in 2004, Facebook has grown from hundreds of users to over 500 million active users (Facebook, 2011). Facebook was originally designed for college staff, but it is now open to anyone thirteen years of age or older. This is because in 2006, Facebook lifted its ".edu requisites," according to which users had to have an email address with an educational suffix (i.e., an email address ending in .edu). With that development, there was a mass movement to Facebook as the social networking portal of choice for most social network users (Mazer et al., 2007, 2009).

The Web 2.0 aspect of Facebook provides a tool for friends to keep in touch and for individuals to have a presence on the Web without needing to build a website. Facebook makes it easy to upload pictures and videos, making its use so simple that nearly anyone can publish a multimedia profile (Mitrano, 2008). Facebook has made it easy to find friends by using one's email address, to search for people by name, or to pull up listings based on a variety of demographic variables. With a public profile on Facebook, a staff member can be found by any of the 500 million other users.

Each Facebook profile has a "wall" where friends can post comments. Since the wall is viewable by all the staff's friends, wall postings are basically a public conversation. By default, staff members can write personal messages on friends' walls or send a person a private message; the latter will show up in their private inbox, similar to an email message. Facebook offers tools to develop and maintain relationships that are of particular importance in emerging adulthood. Recently, the use of messaging via online social networking has surpassed email as the primary means of communication between staff (Lenhart et al., 2010).

Facebook allows each user to set privacy settings. For example, if a staff member has not added a certain person as a friend, the user can have his or her privacy set so that other users will not be able to view the user's profile. A user can adjust the privacy settings to allow other users or peers to view portions or all of the profile. Users can also create a limited profile, which allows users to hide certain parts of the profile from a list of users that an individual selects.

Another feature of Facebook that users like is the ability to add gaming applications to a user page. Facebook applications are programs developed specifically for Facebook profiles. The most popular of these programs include interactive games such as Farm Town, Mafia Wars, and thousands of other interactive multiuser applications. Since most game applications save scores or assets, friends can compete against each other or against millions of other Facebook users.

Providing a cybersecure environment in organizations is ever challenging not only because of the dangers inherent in the Internet and social networking, but also because of the added dimension of staff bringing their personal technology into the organizational environment, where it can access a host of technological resources.

Smart Phones

Staff increasingly utilize their personal technology, most notably smart phones, to access social networking sites. The use of smart phones exceeds the use of traditional computers in accessing the Internet (Albanesius, 2011; Weintraub, 2011). A smart phone is a cellular phone that combines the functions of a networked computing system and a mobile phone. The smart phone hosts a wealth of technology and applications (apps) that are as robust as those of a PC in many situations. In addition to its standard audio and text capabilities, smart phones typically serve as video and still cameras, audio recorders, media players, and mobile computers. They incorporate apps and browsers that can do a number of things, ranging from video conferencing to GPS navigation to social networking via Wi-Fi and broadband Internet protocol network access.

A growing issue with cyberlife and smart phone usage relates to geolocation tools such as Facebook Places. These services have become very popular apps, especially in the social networking environment. The apps allow users to "check in" at locations (e.g., neighborhood businesses) via their mobile phone. Their location is then sent to their friends, in many cases with a map showing their exact location.

Geolocating is used primarily as a marketing tool for businesses, giving reduced-cost or free merchandise for app users who visit their store. However, there are some obvious security considerations. Every time someone checks into a location publicly, they are telling the world exactly where they are. Location sharing could encourage stalking, as well a host of other hazardous issues because the user is broadcasting his or her physical location via the Internet or social media site.

Texting also needs to be considered when looking at smart phones and cyber-security. Technically, texting is not a Web-based technology. However, it has become widely diffused and has become synonymous with modern youth in the proliferation of cell phone technology. As most know, text messaging has become the preferred method of communication for many. While texting, like other cybertechnologies, has many positive benefits for staff, it can also create issues for the security manager.

This section has attempted to provide an understanding of cyber use by staff, discussing both its positive attributes and consequent risks. Providing a technology-secure environment is a very broad topic and includes many aspects of staff life. It is also obvious that technology will continue to play a large part in organizational environments and in the lives of staff members. Therefore, security managers should understand the potential role that organizations can have in facilitating a secure cyberenvironment.

PROVIDING A SECURE CYBERENVIRONMENT

The second part of this chapter addresses the role of providing a cybersecure environment for staff. Providing cybersecurity is contingent upon security managers understanding the environment, risks, and vulnerabilities of technology use. Because of the breadth of the issues and complexity of the environment, there is no more valid approach to achieving a secure cyberenvironment in organizations than to provide policy guidance to organizational stakeholders. This approach will use the security manager to facilitate safe, secure, and ethically sound practices by staff.

It is not the intent of this section to address cybersecurity from the view of a network administrator but to make the security manager aware of what is needed to make an organization's cyberenvironment secure. An organization must have both policies and procedures to provide for authentication, firewalls, and virus protection on its computer systems. Authentication will provide a secure login and allow individuals to access only the data and applications they require. The firewall also serves as an access control device that will limit user access to unwanted programs and data. Virus protection programs will protect the organization from viruses and malware on the network. Each of these three areas must be addressed by computing professionals in that environment. The best application for the security manager is to provide stakeholders with information regarding the risks associated with the cyberenvironment and, subsequently, how to make them safe, secure, and ethical employees.

The task of cybersecurity is daunting because there are so many aspects of a staff member's cyber life that are out of the reach of the organization. Staff members have technology, and they use it very frequently outside the realm of organizational mentorship. They use technology to communicate peer to peer, often without knowledge, concern, or ethical understanding. Therefore, policies, training, and codes of conduct infused with cybersecurity, safety, and ethics are important for providing staff with the ability to make good cyberdecisions. Teaching staffs to be good cybercitizens will better ensure they do not become cybercrime victims or perpetrators in the future. Because technology is diffused at all organizational levels, cybersecurity should be mandated throughout the organization.

One basic principle of cybersecurity is that staff should always be taught to avoid strangers in the cyberworld. Staff must understand that people are not always who they say they are in cyberspace. If a staff member receives emails or other communications that seem suspicious, they should not respond but instead should make security staff aware. Another basic principle that staff should understand is that they should not access non-work-related content on the Internet. Be it audio, video, or some other media, it should be stressed in organizational policy that this is not allowed. Staff must understand the concept of organizational cybercitizenship. At any level of technological interface, staff should be aware that they must be good cybercitizens as an underlying premise.

CYBERSECURITY IN THE ORGANIZATION

Making staff into good cybercitizens is a goal for all organizational members. Computers and cybertechnologies enhance staff performance by engaging, involving, and empowering them, and security management addressing cybersecurity as a part of that relationship is essential. Moreover, it is the responsibility of organizations to provide staff with basic cybersecurity and ethical skills as a portion of their organizational process. Educate staff regarding an acceptable use policy, and include the consequences for violating it. Instill a sense of cyberresponsibility. The role of an organization in facilitating cybercitizenship is going to be even more challenging in the future. The best time to begin establishing a culture of cybersecurity in your organization is the first day you introduce your staff to technology.

EXERCISE

Many employees are using smart phones to access social networking sites while at work. They are using these forums to tease and taunt other employees. This has created a disturbance in the organization.

- Does the issue of cyberthreats differ from other organizational threats?
- What factors must be considered in addressing the issue?
- How should the occupational safety manager address the issue?

REFERENCES

Acar, A. (2008). Antecedents and consequences of online social networking behavior: the case of Facebook. *Journal of Website Promotion*. 3(1–2): 62–83.

Albanesius, C. (2011). Smartphone shipments surpass PCs for first time: what's next? PCMag.com, February 8. Available June 1, 2013, at http://www.pcmag.com/article2/0,2817,2379665,00.asp

Department of Homeland Security. (n.d.). Cybersecurity. Available June 1, 2013, at http://www.dhs.gov/cybersecurity-overview

Facebook Blog. (2011). https://www.facebook.com/blog/. Retrieved December 21, 2011, from https://www.facebook.com/blog/blog.php?post=409753352130

Lenhart, A., Madden, M., Smith, A., Purcell, K., Zickuhr, K., & Rainie, L. (2010). Social Media & Mobile Internet Use Among Teens and Young Adults. Retrieved November 9, 2011, from http://web.pewinternet.org/~/media/Files/Reports/2010/PIP_Social_Media_and_Young_Adults_Report_Final_with_toplines.pdf

Madge, C., Meek, J., Wellens, J., and Hooley, T. (2009). Facebook, social integration and informal learning at university: it is more for socializing and talking to friends about work than for actually doing work. *Learning, Media and Technology*. 34(2): 141–155.

Mazer, J. P., Murphy, R. E., and Simonds, C. J. (2007). I'll see you on "Facebook": the effects of computer-mediated teacher self-disclosure on staff motivation, affective learning, and classroom climate. *Communication Education*. 56(1): 1–17.

Mazer, J. P., Murphy, R. E., and Simonds, C. J. (2009). The effects of teacher self-disclosure via Facebook on teacher credibility. *Learning, Media, & Technology*. 34 (2): 175–183.

Mitrano, T. (2008). Facebook 2.0. *Educause Review*. 43(2): 72.

Rainie, L. (2011). The new education ecology. Sloan Consortium Orlando. Available June 1, 2013, at http://www.slideshare.net/PewInternet/the-new-education-ecology

Shier, M. T. (2005). The way technology changes how we do what we do. *New Directions for Staff Services*. 1(12): 77–87.

Steinfield, C., Ellison, N. B., and Lampe, C. (2008). Social capital, self-esteem, and use of online social network sites: a longitudinal analysis. *Journal of Applied Developmental Psychology*. 29 (6): 434–445.

Subrahmanyam, K., Reich, S. M., Waechter, N., and Espinoza, G. (2008). Online and offline social networks: use of social networking sites by emerging adults. *Journal of Applied Developmental Psychology*. 29(6): 420–433

Weintraub, S. (2011). Smartphones pass PCs in sales. CNN Money. Available June 1, 2013, at http://tech.fortune.cnn.com/2011/02/07/idc-smartphone-shipment-numbers-passed-pc-in-q4-2010/

12 Occupational Safety Investigations

Investigation is the process for obtaining facts and information for a better understanding of an event or to determine the cause of an event. Safety investigations involve fact finding for the purpose of identifying safety breaches, loss, and safety concerns. Compliance is required in 29 CFR §1904.0, "Recording or Reporting a Work-Related Injury, Illness, or Fatality." An investigation may expose activity violating Occupational Safety and Health Administration (OSHA) regulations, company policy, criminal laws, and administrative regulations. For example, an employee falsifying company records may be disciplined by the company and be subject to criminal penalties. The company may face a fine, a lawsuit, or administrative sanctions for the activity.

The Occupational Safety and Health (OSH) Act of 1970 requires certain employers to prepare and maintain records of work-related injuries and illnesses. Furthermore, the investigative process may be used to gather evidence in the event of a lawsuit. For example, an employee driving a company car to the post office collides with another vehicle, causing injury to the driver of the other vehicle. The police collision report is designed to provide a neutral investigation and identify facts as to which party was at fault. The efficacy of the collision report can be biased by a number of factors: the officer omits a fact not important to his or her investigation, evidence may have been moved or overlooked, or information may have been incorrectly recorded. An investigation by company personnel seeks to preserve evidence, statements, and facts to protect against future litigation against the company. As a government agent, this writer investigated a fender bender involving a citizen who was rear-ended by a government vehicle. The "severe" injury to the citizen's neck seemed out of place for the minor damage noted to the citizen's vehicle; the government vehicle was not damaged. When questioned about his injury, the citizen stated that he would likely have to see his brother-in-law the chiropractor for the next two years. Realizing his mistake, he quickly added that he meant the statement as a joke. Nonetheless, the statement was recorded and used in later settlement negotiations.

Investigation may uncover evidence of a criminal nature. As the safety investigation develops, information relating to a criminal act should be reported to the police. Some jurisdictions make it a criminal act in itself not to report a serious crime. 18 U.S.C. §4, "Misprision of a Felony," makes it a crime not to report a federal violation to the authorities. When law enforcement agencies are notified, they will take over the investigation for a legitimate reason. They are responsible for enforcing the laws and must preserve evidence in preparation for criminal legal proceedings. Your cooperation with law enforcement will complement the investigation, but it may also expose the company to civil and even criminal liability. Legal counsel should be informed of any requests for information, especially court orders. Failure to cooperate with a criminal investigation may result in criminal or court sanctions for obstruction of justice.

WHO, WHAT, WHEN, WHERE, HOW, AND WHY

Common questions to be answered are *who*, *what*, *when*, *where*, *how*, and *why*:

Who: Who committed the act? Who reported the event? Who has information on the event? Who witnessed the event?

What: What happened? What proof do we have that the event actually occurred? Is what happened a crime? Did what happen violate a policy? Could what happened subject us to a lawsuit?

When: When did the event occur? How do we know when the event occurred?

Where: Where did the event occur? Was where the event occurred on property owned, leased, or controlled by the company? Does it matter for the company (for liability) where the event occurred?

How: How did this occur? How can we prevent it from happening again? How did we learn about the event? How did we respond to the event?

Why: Why did this occur? Why did they commit the act? Why didn't they report the incident?

Safety investigations can be similar to criminal investigations. The basic questions of who, what, where, when, how, and why provide the foundation for solving the event. The more detailed answers to those questions lead to identifying the problem and finding solutions to alleviate future events.

The investigation gathers evidence. Evidence is any object, statement, or thing that tends to prove or disprove a material fact. The investigator gathers evidence two ways: through personal observation and the observations of others. In most cases, an event requiring further investigation is brought to the attention of the investigator. Discovery may be by safety personnel, other employees, visitors, anonymous tips, and/or surveillance. The event may be observed visually, audibly, or through examination of documents, tests, or measuring devices. Have in mind the objective of your investigation: arrest and prosecution of the culprit, recovery of goods, discovering the leak, and the like. Valuable time can be wasted by going in a direction that is of no value to your objective. You may be more concerned with the recovery of your company's stolen goods, while law enforcement resources may be best used in the prosecution of the criminal act.

On discovery of the event, physical evidence should be preserved and protected from contamination (intentional or unintentional) from bystanders and other curious or interested persons. Safety personnel should remain at the event scene to assure the integrity of the evidence. Searching and recovery of evidence should be conducted immediately, with photographs taken of each item. A diagram of the scene and location of evidence should be completed.

Crime scene diagrams should accurately depict the scene of the event. The room or area and the locations of furniture, evidence, and so on should be measured and recorded to show the relationship of items at a scene. Before any evidence is moved, a photo should be taken, first with the item only, then with an identifying number or mark. This number should track with the evidence throughout the investigation.

Evidence should be stored in a secure area with a notation of the location, date, and who seized the evidence. Examination of evidence should be limited to persons necessary for the investigation, and all persons who have handled the evidence should be noted. Some evidence may need to be examined by a laboratory or other people with specialized experience. For example, financial records on a suspected embezzlement should be examined by a certified public accountant with forensic accounting credentials. Evidence that is moved for examination must identify a chain of custody, again, to maintain the integrity of the evidence. Each person who has "custody" of the item, for any period of time, however brief, should be noted. If the need arises for the evidence to be introduced at trial, the chain of custody will prove that the item introduced at trial is the same item discovered at the event scene. The occupational safety manager should maintain a log to document the chain of custody.

Witnesses should be interviewed and asked to submit a written statement. The basic who, what, when, where, how, and why should be asked to develop facts and generate leads for more information. A search for additional witnesses should fan out from the scene. In addition to questioning people, look for monitoring cameras that may have recorded the event and potential approach and escape routes from the event. In extreme cases, you may need to post signs or advertise for possible witnesses to the event. Rewards may encourage witnesses to come forward. Anonymous witness information can be effective in an investigation. While the witness's credibility cannot be verified, the information given can be corroborated with known information, thereby adding to its credibility. Anonymous sources require further investigation, but can lead to the right person for answers and a solution to your event. Anonymous sources may lead you to an event before something occurs, allowing you to set up personnel or cameras to capture the culprit in the process of the act.

Investigators should be trained in interviewing skills to listen and watch for deceptive or revealing behavior from witnesses. It is not uncommon for the perpetrator to be among those who witnessed the event. It is good to have two investigators during interviews when possible. One investigator can conduct the interview, watch the witness for nonverbal behavior, and listen to the words used, while the second investigator can focus on taking notes.

Interview notes are taken at the time of the interview. The purpose of the notes is to immediately record the witness's statement, paying attention to key facts that may have a bearing on the case. At the scene of an event, investigators work quickly to get an idea of what happened. Notes are often sketchy and written in a modified shorthand or in acronyms the interviewer will understand when he or she prepares a report of the interview.

When a formal statement or report of the interview is written, it is possible for interviewer bias to omit a point that does not support their position. If the formal statement is challenged, the notes are used to clarify the statement. Since the notes were prepared at the time of the interview, which is closer in time to the event than when the report of the interview was prepared, they are more likely to accurately reflect the event.

Interviews of potential subjects should always be conducted with two investigators, primarily for the reasons listed here. Two investigators complement each other; for example, one investigator may think of a question another didn't think of. Subject interviews are usually conducted in an office or interview room. If possible, the subject interview should occur in the safety office. The room should be free from distractions. A picture on a wall can give the interviewee something to focus on. Chairs should be positioned with no table between the investigators and the interviewee. This allows you to watch for body language. This writer liked to place the interviewee in a swivel chair. The chair would telegraph movements, which can indicate nervousness, deception, or both.

At the conclusion of the interview, a statement should be prepared by the investigators for the interviewee to sign, especially when the interviewee confesses to the act or makes admissions indicating some culpability in the matter. A confession is an admission of guilt that accounts for all of the elements of a crime or civil offense. Admissions are when a person admits to an act, or part of an act, but not culpability for a crime or civil offense. Admissions can be used to prove guilt (e.g., a person may admit they took a computer from the supply room, but deny that they stole it).

The investigator should prepare the statement rather than have the interviewee write out a personal statement. The interviewee may intentionally or inadvertently leave out key information. The interviewee is given the statement and time to read the statement and make corrections before acknowledging the statement as a true reflection of his or her comments. Interview notes are always retained.

All evidence, records, and statements should be stored and recorded in an investigation file. A filing system should be established so files can be found and accessed efficiently. The file should identify the location of evidence and list the status of the case. An investigative report should summarize all of the evidence and list potential witnesses and their potential testimony. A complete history of suspects should be included as well.

INVESTIGATIVE REPORT

When all of the evidence has been collected and examined, statements have been taken, and reports have been received, the case file should be reviewed and material prepared into a report. See the "Format for Investigative Reports" section for a sample outline of what is needed. The investigative report begins with a statement of the event and the persons responsible or liable for the event. Provide a synopsis of the matter so that the reader can get a general idea of what happened and what action you are proposing. Your action may suggest criminal prosecution, civil remedies that may include employee disciplinary action, changes to policy, or preparation for civil action against the organization.

A table of contents page helps the reader move quickly to key areas in the report. The appendix will contain detailed reports and other documents that are useful for prosecution or civil action, or defense from prosecution or civil action. In between the table of contents and appendix are summaries of the key evidence.

Since this document may go outside the organization and even become public in a trial, trade secrets and others should be noted but not included in your report. The caution here is not to try to conceal or misinterpret evidence. All evidence should be noted with confidential material being identified as such. Attorneys will deal with disclosure in deciding to prosecute a matter.

FORMAT FOR INVESTIGATIVE REPORT

- Cover page
  - Title of investigation
  - Case number
  - Name of subjects
  - List of violations
  - A synopsis of acts and events
- Table of contents (for headings delineated below)
- Criminal, civil, and administrative history
  - Related proceedings
  - Contact persons
- Potential witnesses
  - Name of witness and what they will testify to
  - Page number where the witness statement will be found
- List of documents
- List of evidence
  - Who will testify to what piece of evidence
  - Location of evidence (police storage, company vault, etc.)
- Appendix
  - Documents, statements, and reports
  - Case studies
  - Filing safety reports required by compliance, such as §1904.0, does not mean that the employer or employee was at fault, that an OSHA rule has been violated, or that the employee is eligible for workers' compensation or other benefits.

EXERCISE

In recent years, a number of companies have been cited for not maintaining injury records. What would you do as an occupational safety manager to ensure that this does not happen?

13 Safety and Security Management for Chemical Facilities

Chemicals pose a great risk to many organizations and a challenge to safety and security managers. The concern of this chapter is not just with organizations that produce or store large chemical stockpiles, but also with those facilities that use chemicals that are secondary to their primary mission. The focus of this chapter is going to be on the impact that the Department of Homeland Security (DHS) has had on the regulation and security required by those organizations that use and store chemical inventories. This chapter relies heavily on DHS documents to try to simplify the chemical regulation process and the subsequent role for security management. The DHS provided its "Chemical Facility Anti-Terrorism Standards" as a framework for securing chemicals from the possibility of harm or theft by terrorist groups. The DHS has the authority to regulate chemical facilities that present high levels of security risk.

On October 4, 2006, President Bush signed the Department of Homeland Security Appropriations Act of 2007, which provides the DHS with the authority to regulate the security of chemical facilities. The Chemical Facility Anti-Terrorism Standards (CFATS; 6 CFR §27) Interim Final Rule (IFR) was published on April 9, 2007. The purpose of 6 CFR §27 is to lower the risk posed by certain chemical facilities. CFATS requires chemical facilities to provide the DHS with information to determine whether they are a covered facility and are required to meet certain security performance requirements. In order to identify high-risk chemical facilities, the DHS has identified chemicals for preliminary screening based on the belief that such chemicals, if released, stolen, diverted, and/or contaminated, have the potential to create significant human health and/or life consequences.

If a facility possesses a chemical that is on the DHS chemicals of interest list in 6 CFR §27, Appendix A (DHS, 2007a), at or above the screening threshold quantity for any applicable security issue, the facility must complete and submit a chemical security assessment, known as the Chemical Security Assessment Tool (CSAT) Top-Screen, to the DHS. The CSAT is an easy-to-use online questionnaire that must be completed by facilities that possess any chemical on the chemicals of interest list at or above the listed quantity for each chemical. 6 CFR §27.200 of CFATS authorizes the DHS to collect information from chemical facilities on a broad range of topics related to the potential consequences of or vulnerabilities to an attack or incident. The Top-Screen is one method that the DHS may use to gather such information. After analyzing a facility's Top-Screen information, the DHS will make a preliminary determination of whether a facility presents a high level of security risk and therefore must comply with additional requirements of CFATS. (See Appendix C of this volume.)

The number of organizations that have to adhere to this standard is extensive. The DHS has attempted to identify facilities covered by its chemical security regulation. Facilities that may be required to comply with at least some provisions of the CFATS regulation will largely fall into the following categories:

- Chemical manufacturing, storage, and distribution
- Energy and utilities
- Agriculture and food
- Paints and coatings
- Explosives
- Mining
- Electronics
- Plastics
- Healthcare (DHS, 2007c)

To determine which chemical facilities meet the CFATS criteria for high-risk chemical facilities, the DHS developed the Top-Screen. Exemptions to CFATS regulations are as follows:

- Facilities regulated pursuant to the Maritime Transportation Safety Act (MTSA)
- Public water systems, as defined in the Safe Drinking Water Act
- Water treatment facilities, as defined in the Federal Water Pollution Control Act
- Facilities owned or operated by the Department of Defense or the Department of Energy
- Facilities subject to regulation by the Nuclear Regulatory Commission (DHS, 2007c)

The DHS does not currently regulate railroad facilities that store in rail cars large quantities of chemicals or materials on the DHS chemical of interest list, and it does not request that railroads complete the CSAT Top-Screen. Likewise, the DHS has no intention at this time of requiring long-haul natural gas pipelines to complete the CSAT Top-Screen; however, chemical facilities otherwise covered by this regulation and with a pipeline within their boundaries must identify the pipeline as an asset and address it, as appropriate, in a site security plan (DHS, 2007c).

Under CFATS, Congress directed the DHS to identify and secure those chemical facilities that present the greatest security risk. From their definition, *security risk* is a function of the following:

- The consequence of a successful attack on a facility
- The likelihood that an attack on a facility will be successful (vulnerability)
- The intent and capability of an adversary in respect to attacking a facility (threat)

Therefore, Congress and the federal administration have directed the DHS to ensure the security of specifically high-risk chemical facilities.

RISK-BASED PERFORMANCE STANDARDS

Since each chemical facility faces different security challenges, Congress explicitly directed the DHS to issue regulations establishing risk-based performance standards for securing chemical facilities. Performance standards are particularly appropriate in a security context because they provide individual facilities with the flexibility to address their unique security challenges. Using performance standards rather than prescriptive standards also helps to increase the overall security of the sector by varying the security practices used by different chemical facilities. Security measures that differ from facility to facility mean that each presents a new and unique problem for an adversary to solve.

RISK-BASED TIERING

The DHS has developed a risk-based tiering structure that will allow it to focus resources on the high-risk chemical facilities. To that end, the DHS will assign facilities to one of four risk-based tiers ranging from high (Tier 1) to low (Tier 4) risk (DHS, 2007c).

Assignment of tiers is based on an assessment of the potential consequences of a successful attack on assets associated with chemicals of interest. The DHS uses information submitted by facilities through the Top-Screen and security vulnerability assessment processes to identify a facility's risk, which is a function of the potential impacts of an attack, the likelihood that an attack on the facility would be successful, and the likelihood that such an attack would occur at the facility.

Facilities that complete the CSAT Top-Screen and do not meet the consequence thresholds do not need to comply with CFATS. The DHS recognizes that facilities have dedicated and invested time, resources, and capital to identify vulnerabilities and improve overall security. Facilities will be able to make use of information from these improvements. Facilities may also leverage their existing security measures to work toward compliance with CFATS, specifically the risk-based performance standards.

The DHS considers a variety of factors in determining the appropriate tier for each high-risk facility, including information about the public health and safety risk, as well as the presence of chemicals with a critical impact on the governance mission and the economy.

The security measures needed to satisfy the risk-based performance standards for each covered facility correspond to the security risks presented by the facility. Accordingly, facilities that present a higher risk will be required to meet more rigorous risk-based performance standards.

TOP-SCREEN PROCESS

The first step for a facility to take in determining whether it is covered under CFATS is to review the exemptions. Unless a facility is exempt, it must conduct a review to determine whether it possesses any chemicals of interest at or above the listed screening threshold quantities. The DHS has listed the security issue(s) associated with each chemical of interest; each chemical presents at least one security issue, and some chemicals present multiple security issues. Where there are multiple security issues associated with a chemical, a facility must complete and submit a Top-Screen if it meets or exceeds the screening threshold quantities for any of the applicable security issues. If a facility determines that it possesses a chemical of interest at or above any applicable screening threshold quantities, the facility must register with the DHS for access to DHS CSAT1 and complete the Top-Screen Survey Application. Using the information submitted via the Top-Screen Survey Application, the DHS will make a preliminary determination of whether the facility presents a high level of security risk.

The CSAT Top-Screen follows a logical, two-step data collection process (DHS, 2007b):

1. Step 1 involves collecting basic facility identification information.
2. Step 2 involves collecting information about the chemicals that a facility possesses, manufactures, processes, uses, stores, and/or distributes.

Questions cover the following security issues:

- Release-toxic, release-flammable, and release-explosive chemicals with the potential for impacts within and beyond a facility
- Theft of explosive or improvised explosive device precursor (theft or diversion—EXP–EDP) chemicals, theft of weapons of mass effect (theft or diversion—WME) chemicals, and theft of chemical weapon or chemical weapon precursor (theft or diversion—CW–CWP) chemicals
- Sabotage or contamination of chemicals
- Chemicals that are critical to the government's mission and the national economy

The Top-Screen also provides additional questions for each of the facility types:

- Petroleum-refining facilities
- Liquefied natural gas storage facilities (DHS, 2007b)

Upon completion of the Top-Screen, facilities will be presented with one of two outcomes: a preliminary determination that a facility is not high risk or that a facility is high risk. The DHS will notify the facility of:

- Its preliminary status as a high-risk facility
- Its preliminary placement in a risk-based tier
- The specific chemical(s) of interest and related security issues that need further analysis

SECURITY VULNERABILITY ASSESSMENTS

If the DHS makes a determination that a facility is high risk, they will require the facility to complete a security vulnerability assessment to identify and assess the security of a facility's critical assets in light of the security issues raised by its procession of the subject chemicals.

Following a facility's submission of the security vulnerability assessment and its analysis by the DHS, the DHS will either confirm that a facility is high risk or inform a facility that the DHS no longer considers the facility to be high risk or subject to further regulation under CFATS. For facilities confirmed to be high risk, the DHS will communicate the final facility tier determination, and the facilities must develop and implement site security plans that satisfy the risk-based performance standards enumerated in 6 CFR §27.230.

SITE SECURITY PLANS

The site security plan must meet the following standards:

1. Address each vulnerability identified in the facility's security vulnerability assessment, and identify and describe the security measures to address each such vulnerability.
2. Identify and describe how security measures selected by the facility will address the applicable risk-based performance standards and potential modes of terrorist attack, including, as applicable, vehicle-borne explosive devices, waterborne explosive devices, ground assault, or other modes or potential modes identified by the DHS.
3. Identify and describe how security measures selected and utilized by the facility will meet or exceed each applicable performance standard for the appropriate risk-based tier for the facility.
4. Specify other information that the assistant secretary deems necessary regarding chemical facility security. A covered facility must complete the site security plan through the CSAT process, or through any other methodology or process identified or issued by the assistant secretary. Covered facilities must submit a site security plan to the DHS in accordance with the schedule provided in §27.210. When a covered facility updates, revises, or otherwise alters its security vulnerability assessment pursuant to §27.215(d), the covered facility shall make corresponding changes to its site security plan. A covered facility must also update and revise its site security plan in accordance with the schedule in §27.210. A covered facility must conduct an annual audit of its compliance with its site security plan.

RISK-BASED PERFORMANCE STANDARDS

Covered facilities must satisfy the performance standards identified in this section. The assistant secretary will issue guidance on the application of these standards to the risk-based tiers of covered facilities, and the acceptable layering of measures used to meet these standards will vary by risk-based tier. Each covered facility must select, develop in their site security plan, and implement appropriately risk-based measures designed to satisfy the following performance standards:

1. *Restrict the area perimeter.* Secure and monitor the perimeter of the facility.
2. *Secure site assets.* Secure and monitor restricted areas or potentially critical targets within the facility.
3. *Screen and control access.* Control access to the facility and to restricted areas within the facility by screening and/or inspecting individuals and vehicles as they enter, including

   - Measures to deter the unauthorized introduction of dangerous substances and devices that may facilitate an attack or actions having serious negative consequences for the population surrounding the facility
   - Measures implementing a regularly updated identification system that checks the identification of facility personnel and other persons seeking access to the facility and that discourages abuse through established disciplinary measures

4. *Deter, detect, and delay.* Deter, detect, and delay an attack, creating sufficient time between the detection of an attack and the point at which the attack becomes successful, including measures to

   - Deter vehicles from penetrating the facility perimeter, gaining unauthorized access to restricted areas, or otherwise presenting a hazard to potentially critical targets.
   - Deter attacks through visible, professional, well-maintained security measures and systems, including security personnel, detection systems, barriers and barricades, and hardened or reduced value targets.
   - Detect attacks at early stages through countersurveillance, frustration of opportunity to observe potential targets, surveillance and sensing systems, and barriers and barricades.
   - Delay an attack for a sufficient period of time to allow appropriate response through onsite security response, barriers and barricades, hardened targets, and well-coordinated response planning.

5. *Shipping, receipt, and storage.* Secure and monitor the shipping, receipt, and storage of hazardous materials for the facility.
6. *Theft and diversion.* Deter the theft or diversion of potentially dangerous chemicals.
7. *Sabotage.* Deter insider sabotage.
8. *Cybersecurity.* Deter cybersabotage, including by preventing unauthorized onsite or remote access to critical process controls, such as supervisory control and data acquisition (SCADA) systems, distributed control systems (DCS), process control systems (PCS), industrial control systems (ICS), critical business systems, and other sensitive computerized systems.
9. *Response.* Develop and exercise an emergency plan to respond to security incidents internally and with the assistance of local law enforcement and first responders.

10. *Monitoring.* Maintain effective monitoring, communications, and warning systems, including

- Measures designed to ensure that security systems and equipment are in good working order and are inspected, tested, calibrated, and otherwise maintained
- Measures designed to regularly test security systems, note deficiencies, correct for detected deficiencies, and record results so that they are available for inspection by the DHS
- Measures to allow the facility to promptly identify and respond to security system and equipment failures or malfunctions

11. *Training.* Ensure proper security training, exercises, and drills of facility personnel.
12. *Personnel surety.* Perform appropriate background checks on and ensure appropriate credentials for facility personnel and, as appropriate, for unescorted visitors with access to restricted areas or critical assets, including

- Measures designed to verify and validate identity
- Measures designed to check criminal history
- Measures designed to verify and validate legal authorization to work
- Measures designed to identify people with terrorist ties

13. *Elevated threats.* Escalate the level of protective measures for periods of elevated threat.
14. *Specific threats, vulnerabilities, or risks.* Address specific threats, vulnerabilities, or risks identified by the assistant secretary for the particular facility at issue.
15. *Reporting of significant security incidents.* Report significant security incidents to the DHS and to local law enforcement officials.
16. *Significant security incidents and suspicious activities.* Identify, investigate, report, and maintain records of significant security incidents and suspicious activities in or near the site.
17. *Officials and organizations.* Establish official(s) and an organization responsible for security and for compliance with these standards.
18. *Records.* Maintain appropriate records.
19. *Additional standards.* Address any additional performance standards the assistant secretary may specify (Department of Homeland Security, 2007b).

Since 9/11, the security of chemicals has become highly regulated by the DHS. Chemicals pose a great challenge to security managers. The concern of this chapter is not just with organizations that produce or store large chemical stockpiles, but also with those facilities that use chemicals that are secondary to their primary mission. The impact that the DHS has on the regulation and security required by those organizations that use and store chemical inventories is astounding.

EXERCISE

The DHS and the CFATS regulations are concerned with mandatory safety and security requirement for organizations who use and store chemicals. As part of this process they must maintain standards of security in protecting those chemical assets. Provide a detailed list of those safety and security requirements.

REFERENCES

Department of Homeland Security. (2007a). 6 CFR §27: Appendix to chemical facility anti-terrorism standards; final rule. Available June 5, 2013, at http://www.dhs.gov/xlibrary/assets/chemsec_appendixa-chemicalofinterestlist.pdf
Department of Homeland Security. (2007b). Chemical facility anti-terrorism standards. Available June 5, 2013, at http://www.dhs.gov/chemical-facility-anti-terrorism-standards
Department of Homeland Security. (2007c). Identifying facilities covered by the chemical security regulation. Available June 5, 2013, at http://www.dhs.gov/identifying-facilities-covered-chemical-security-regulation

14 Safety and Security Management for the Energy Sector

Regarding safety and security, organizations that are in the energy sector are susceptible to great risks, and this poses great challenges to those responsible for minimizing risks. The concerns of this chapter are not just with organizations that produce or store energy in their primary mission, but also on the impact that the Department of Homeland Security (DHS) has had on the safety regulation and security required by those organizations that make up the energy sector.

The energy sector represents a very important component of our nation's critical infrastructure. More than 80% of the country's energy infrastructure is owned by the private sector, and it is integral to growth and production across the nation (National Infrastructure Protection Plan [NIPP], 2011). The energy infrastructure is divided into three interrelated segments: electricity, petroleum, and natural gas. The U.S. electricity segment contains more than 6,413 power plants, including 3,273 traditional electric utilities and 1,738 nonutility power producers. Approximately 48 percent of electricity is produced by combusting coal, which is primarily transported by rail; 20% is produced in nuclear power plants; and 22% by combusting natural gas (DHS, 2013).

The heavy reliance on pipelines to distribute products across the nation highlights the interdependencies between the energy and transportation sectors. The reliance of virtually all industries on electric power and fuels means that all sectors have some reliance on a secure and stable energy sector. Many energy sector stakeholders and security managers have extensive experience with infrastructure protection and have more recently focused their attention on cybersecurity.

PROTECTING THE ENERGY SECTOR

The energy sector is very vulnerable. In 2008, an analyst at the Central Intelligence Agency stated publicly that cyberattacks had already been used to disrupt electrical power in multiple cities outside the United States (McMillan, 2008). This alarming report and others, however, fail to describe the extent to which a complex system such as a national electric grid could be vulnerable to cyberattacks.

Demonstrations of a cyberattack were exemplified against control systems that regulate an electric generator in Idaho. In this demonstration, under the code name "Aurora," the Department of Energy's Idaho National Laboratory manipulated the generator's controls to exploit system weaknesses, which caused the generator to fail. In particular, the attack caused extreme vibrations, which in turn physically destroyed internal components and ultimately caused the generator to catch fire (Meserve, 2007). This kind of attack, which was demonstrated to show feasibility, is likely to be even more effective in much larger generators, such as those in big dams and many coal-fired power plants.

The huge task of protecting the energy sector's organizations is much like security in other organizations. By using a methodological process, the security manager must move to understand the environment, including its assets and their vulnerabilities and criticalities, to formulate a means for protection.

IDENTIFYING ORGANIZATIONAL ASSETS AND GOALS

The first step of the energy sector security management process is to identify assets and specific overall goals for the organization. This part of the planning process should include a detailed overview of each organizational asset and its relationship to goals. Where possible, objectives should be described in quantitative or qualitative terms. These quantitative and qualitative goals should be measurable. Being able to measure needs, as well as outcomes, is fundamental to security management as well as to the organization.

The application of energy sector security programs shall follow the critical thinking process regardless of the asset. The asset will have a direct impact on the application of the process and how countermeasures are to be implemented. Application of this security standard ensures a comprehensive approach to meeting organizational security needs in a threat environment, and ensures that the scope of security is commensurate with the risk posed to an asset, relative to cost.

The energy sector has identified six general asset or system characteristics that are important parameters for evaluating the vulnerabilities of the energy sector infrastructure and developing risk management programs.

- *Physical and location attributes*: This attribute relates to the physical location of the asset and the role that geographic location plays in the asset's inherent presence.
- *Cyberattributes*: This considers the impact that cybersystems play in relationship to the asset.
- *Volumetric or throughput attributes*: This attribute is related to the capacity usage of the asset.
- *Temporal and load profile attributes*: This attribute relates to the varying loads that the system may utilize based on, for example, seasonal energy usage.
- *Human attributes*: This relates to the technical knowledge of the energy sector's workforce in maintaining the sector's safety, reliability, and security.
- *Importance of the asset or system to the energy network*: This relates to the interdependency and interrelationships between the larger energy grids (NIPP, 2011).

The security manager must understand the relationship of assets in organizational structures and functions. This includes the physical and logical relationships that assets have with each other in the organizational environment. It can also relate to the impact that one asset can have on another in a loss event. An essential tool for a better understanding the relationship between organizational assets is through the configuration of an asset hierarchy, because of the dependencies on and interdependencies among energy sector infrastructures.

ANALYZING RISKS TO ORGANIZATIONAL ASSETS

An effective security management system demonstrates a careful evaluation of how much security is needed to protect organizational assets. Security managers must realize that too little security means that organizational assets can easily be compromised, while too much security can make assets hard to use or so degraded that performance is negatively affected. Security must be inversely proportioned to an asset's utility. It is given that there is always going to be risk associated with assets and activities. The only way to completely eliminate the risk would, in many cases, make that asset inoperable. Therefore, the role of security management is to find the optimal relationship between organizational processes, assets, and functionality. Although the risk assessment process is more deeply analyzed in Chapter 9, it still maintains merit when looking at the overall role of building a security management program.

If an organizational asset is viewed through the inventory of potential loss events, the security manager must recognize that findings are not necessarily all-inclusive. For each undesirable event where the assessed risk is either less than or exceeds the baseline level of protection, the security manager must identify the countermeasures that will provide a level of protection equivalent to the level of risk. For lower lever risk, minimum countermeasures are typically less stringent, but they may also be less effective in mitigating higher risks, while at the other extreme very high countermeasures are typically more stringent and generally more effective.

As defined in the NIPP (2011) Base Plan, risk is a measure of potential harm that encompasses threat, vulnerability, and consequence. That is, an asset's risk is a function of the likely consequences (C) of a disruption or successful attack; the likelihood of a disruption or attack on the asset, often referred to as the threat (T) to the asset or the asset's attractiveness; and the asset's vulnerability (V) to a disruption or attack. As discussed in this chapter, the energy sector uses a variety of approaches that apply this widely accepted risk management principle to assess risk.

Security and risk managers in the energy sector has extensive experience in the development and application of methodologies for assessing facility and system risk and for prioritizing assets to be protected. Such methodologies have been developed by a variety of sector security partners, including individual energy companies that own and operate energy sector assets, professional and trade associations, academic institutions, research centers, and the Department of Energy as an integral part of meeting its long-standing responsibilities for safety and security and implementing its Conservation Improvement Program (CIP) program for the energy sector.

Because of the diversity of assets in the energy sector, many risk assessment methodologies are used. Some methodologies are tailored to a specific segment of the sector (e.g., electricity, oil, natural gas, or their system components), while others are used to assess risks at the system or sector level.

The Department of Energy, in cooperation with sector security partners, has undertaken programs to assess the risks of key energy infrastructure assets and to provide technology, tools, and expertise to other federal organizations, as well as state and local organizations and the private sector.

These programs have involved establishing partnerships with infrastructure owners and operators, state and local governments, and a wide range of industry associations. Products include vulnerability-related and risk assessment–related methodologies, checklists, lessons learned, support for policy analysis, and guidelines for various types of assets. The energy sector also has worked closely with the DHS in developing and transferring risk assessment methodologies. The sector has participated in the DHS's Buffer Zone Protection Program (BZPP) and has worked with it to develop Risk Analysis and Management for Critical Asset Protection (RAMCAP) modules for petroleum-refining and liquefied natural gas facilities. Given the diversity of facilities in the energy sector and the wide range of methodologies being used successfully to assess risk, a "one-size-fits-all" risk assessment solution is not appropriate.

The energy sector will consider such criteria through the Critical Infrastructure Protection Advisory Council (CIPAC) as the sector evaluates how best to move forward in terms of vulnerability and/or risk assessments that will support the DHS's national risk analysis goals, and how best to improve these methodologies.

The desired levels of protection of assets should be critically determined via a risk-based analytical process or risk assessment. The process will focus on risk as a measurement of potential harm or loss from an undesirable event. Understanding risk means understanding threats, vulnerabilities, and consequences. The level of risk is the combined measure of threats, vulnerabilities, and consequences posed to assets from specified loss events.

If the existing level of protection is insufficient, a determination must be made as to whether the necessary level of protection can be achieved; specifically, if the countermeasure can be physically implemented, and whether the investment is cost-effective. Cost-effectiveness is based on the investment in the countermeasure versus the value of the asset. In some cases, investment in an expensive countermeasure may not be advisable because the life cycle of the asset is almost expired. Additionally, consideration should be given to whether other countermeasures may take priority for funding. Note that *cost-effective* is a different value from *cost-prohibitive*. A countermeasure is cost-prohibitive if its cost exceeds available funding. Funding may exist for a countermeasure, but it may not be a sound financial decision to expend that money for little gain, making it not cost-effective.

ASSESSING CONSEQUENCES

The potential physical and cyber-related consequences of any incident, including terrorist attacks and natural or human-made disasters, are the first factors to be considered in risk assessment. In the context of the NIPP Base Plan, consequence is measured as the range of loss or damage that can be expected. The consequences that are considered for the national-level comparative risk assessment are based on the criteria set forth in Homeland Security Presidential Directive No. 7 of 2003 (HSPD-7). These criteria can be divided into four main categories (NIPP, 2011):

- *Human impact*: Effects on human life and physical well-being (e.g., fatalities and injuries)
- *Economic impact*: Direct and indirect effects on the economy (e.g., costs resulting from disruption of products or services, costs to respond to and recover from the disruption, costs to rebuild the asset, and long-term costs due to environmental damage)
- *Impact on public confidence*: Effects on public morale and on confidence in national economic and political institutions
- *Impact on government capability*: Effects on the government's ability to maintain order, deliver minimum essential public services, ensure public health and safety, and carry out national security–related missions. For example, one DOE power marketing administrations (PMA), the Bonneville Power Administration (BPA), has used the following screening criteria to identify its most critical facilities: economic security, national security, public health and safety, generation, and regional and national grid reliability.

An assessment of all categories of consequence may be beyond the capabilities available for a given risk analysis. Most energy sector assets are not associated with mass casualties, but they may have economic and long-term health and safety implications if disrupted. However, the redundancy of system-critical facilities and the overall system resilience minimize the potential for such consequences.

The complexity, diversity, and interconnectedness of the energy sector dictate the need for assessing consequences at many different levels of detail:

- Asset or facility level
- System, sector, and urban-area level
- Regional and/or national level

These interdependencies may have national, regional, state, and/or local implications and are considered to be an essential element of a comprehensive examination of physical and cyber-related vulnerabilities.

ASSESSING THREATS

The energy sector views threat analysis broadly, to encompass natural events, criminal acts, insider threats, and foreign and domestic terrorism. Natural events are typically addressed as part of emergency response and business continuity planning.

In the context of risk assessment, the threat component of risk analysis is calculated based on the likelihood that an asset will be disrupted or attacked. Such information is essential for conducting meaningful vulnerability and risk assessments. Therefore, the energy sector strongly believes that relevant and timely threat information must be disseminated whenever possible. A number of sector representatives hold national security clearances that facilitate the sharing of classified threat information.

The DHS Homeland Infrastructure Threat and Risk Analysis Center (HITRAC), which conducts integrated threat analysis for all critical infrastructure and key resources (CI/KR) sectors, will work in partnership with owners and operators and other federal, state, and local government agencies to ensure that suitable threat information is made available. Furthermore, the same level of partnership must exist within all levels of federal, state, and local law enforcement.

The following types of threat products provided by HITRAC are needed for the energy sector:

- *Common threat scenarios*, which present methods and tactics that could be employed in attacks against the U.S. infrastructure
- *General threat environment assessments*, which are sector-specific threat products that include known terrorist threat information and long-term strategic assessments and trend analyses of the evolving threats to the sector's critical infrastructure
- *Specific threat information*, which is CI-specific information based on real-time intelligence, and which will drive short-term measures to mitigate risk

More specifically, they help energy facilities, local law enforcement, and others to be more aware of potential indicators of terrorist and/or criminal activity.

ASSESSING VULNERABILITIES

Vulnerabilities are the characteristics of an asset's, system's, or network's design, location, security posture, process, or operation that render it susceptible to destruction, incapacitation, or exploitation by mechanical failures, natural hazards, or terrorist attacks or other malicious acts. Vulnerability assessments identify areas of weakness that could result in consequences of concern, taking into account intrinsic structural weaknesses, protective measures, resiliency, and redundancies.

Historically, the energy sector has been proactive in developing and applying vulnerability assessment methodologies that are tailored to its assets and systems. However, no single vulnerability tool or assessment methodology is universally applicable. Individual energy companies use assessment tools that are developed by professional and trade associations, federal organizations, government laboratories, and private sector firms. The number of tools in use is large, and the vast majority of significant facilities in the energy sector have already undergone assessments using one or more of these tools.

PRIORITIZING INFRASTRUCTURE

The energy sector is characterized by large networks as opposed to discrete assets. These networks are designed to operate with certain levels of reliability, even if portions of them (discrete components, or assets) are out of service. The importance of many of the individual components in the network is highly variable, depending upon the location, time of day, day of the week, month of the year, and many other variables. What might be a critical asset on a Monday morning in January may not be critical on a Saturday afternoon in May.

Owners and operators of energy sector assets and networks have screening processes to identify internal priorities related to business conditions and supply and network reliability to help them ensure continuity of operations. From a grid perspective, the nation's oil and natural gas pipeline systems and electrical grid are designed and operated with built-in redundancy to ensure a certain degree of reliability and resiliency. Industry planning criteria assume that a local grid area can be operated even if one asset is out of service. In addition, during unforeseen events, the industry provides mutual aid to assist in emergency response and prompt restoration (see Chapter 5).

Regional planning groups for the oil and natural gas industry, and historically the Nuclear Energy Regulatory Commission and regional reliability councils for the electricity industry, continuously evaluate network reliability. Their functions are well developed and understood, and the effectiveness of mutual aid agreements can be significantly affected by the nature of an event. Mutual aid partners could also be impacted by an event, and a utility might have to go outside the region to obtain aid.

The Department of Energy will continue to work in partnership with energy sector security partners to evaluate and support existing protective programs and to develop and support new programs that effectively reduce the vulnerability of critical energy assets. The overall strategy will focus on efforts that support the sector's goals to ensure continuity of energy services and business through reliable information sharing, effective physical security and cybersecurity protection, and coordinated response capabilities.

The cornerstone of the overall strategy is partnership with all key stakeholders in the public and private sectors. This approach will continue to take full advantage of the extensive experience and expertise of sector partners and will ensure that repercussions of planned activities are carefully considered. This chapter outlines the methods that energy sector partners will use to assess, select, and implement cost-effective infrastructure protective programs, and it highlights some of the existing cooperative efforts within the energy sector.

DEVELOPING AND IMPLEMENTING PROTECTIVE PROGRAMS

Sector-specific plans should identify long-term technological solutions for protecting physical assets, energy control systems, and related cyber systems.

Some activities in different phases may proceed simultaneously, where feasible, to expedite improvements in a CIP.

Throughout the process, DOE will continue to work with security partners within the framework of the energy sector's goals that support its vision of a "robust, resilient" energy infrastructure in which continuity of business and services is maintained through secure and reliable information sharing, effective risk management programs, coordinated response capabilities, and trusted relationships between public and private security partners at all levels of industry and government (Energy, 2007).

IMPLEMENTING THE SECURITY PROGRAMS FOR SAFETY

When safety managers complete the critical analysis via asset criticality and vulnerability studies, risk assessments, and cost–benefit analysis, then they may implement the security measures that have been determined to best fit the asset risk.

Implementation of new security programs are best accomplished through stages to make it easier for the organization to adapt to changes in the working environment. The security manager and organizational management should understand that there may be user resistance to security functions. It is recommended that staged implementation be performed, starting with the most critical or vulnerable assets.

MONITORING FOR COMPLIANCE

Effective safety management depends on adequate compliance monitoring. Most often, violations of security practices, whether intentional or unintentional, become more frequent and serious if not detected and acted on. Compliance monitoring has two primary activities: detecting security violations and responding to them.

The safety manager should document the response to violations and follow up immediately after noncompliance is detected. The organization should have a designated response group to deal with security violations. Members of the response group should have access to organizational leadership so that severe situations can be dealt with effectively.

A critical part of noncompliance should be the generation of reports for organizational leadership that discuss security violations. An additional objective of monitoring security measures for noncompliance is to identify potential security violations before they dilute the effectiveness of the program or before they cause serious damage.

REEVALUATING ASSETS AND RISKS

Safety management is a discipline that should be dynamic. As changes in the organization or assets occur, a reassessment of the security measures should also occur. Organizational leadership should keep security management abreast of larger changes in the organization so that security operations and measures are prepared to meet those challenges. The importance of sector assets is impacted by changing threats and continually changing consequences. Prioritization in the energy sector is dynamic—it changes constantly and goes on continuously. Static prioritization of energy sector assets could lead to critical decision making based on outdated or erroneous asset information in efforts to direct scarce resources to those assets, systems, and networks that may be the most critical at any point in time. The DOE works with the DHS to identify gaps in existing energy information and to identify publicly available databases or sources that could provide data to support DHS efforts to prioritize assets.

Some DHS, DOE, and other government programs need to allocate resources based on their prioritization (e.g., the DHS's Buffer Zone Protection Program), site assistance visits and comprehensive reviews, as well as state and local initiatives. These programs supplement and support industry efforts. State and local efforts under the NIPP will be based on some measure of the relative importance, risk consequence, and vulnerability of the critical infrastructures within their jurisdictions. This will require that they work with the energy sectors in their jurisdictions so as to understand the importance of critical facilities. In addition, they will need to address policy, regulatory, or other barriers to undertake needed measures and to allow for recovery of prudently incurred costs for those utilities subject to rate regulation. In addition, the DHS is providing funding to state and local entities based on risk assessments of critical infrastructures. The National Asset Database (NADB) is also organized by criteria that may not fully capture the relative importance of energy infrastructure from a systems perspective (NIPP, 2011).

Developing a security management program requires a broad field of knowledge in asset loss prevention, physical security, occupational safety, and intangible asset protection functions. It requires a comprehensive knowledge of organizational assets and the development and implementation of physical measures, policies, procedures, and guidelines to protect those assets.

Security management requires critical thinking skills for developing mechanisms to protect organizational assets. The process of security management utilizes processes of critical thinking, providing a basis for a comprehensive security management program. The dynamic nature of organizations and environments requires that the security response also be dynamic.

EXERCISE

The energy sector is very complex regarding safety and security because it has relationships with multiple organizations throughout the energy sector. Explain how the complexity is a result of interrelationships and how losses in one area can affect other areas.

REFERENCES

Department of Energy. (2007, May). Energy: Critical Infrastructure and Key Resources Sector-Specific Plan as Input to the National Infrastructure Protection Plan. Washington, DC: Department of Energy. http://www.naruc.org/Publications/Energy_SSP_Public3. pdf (accessed January 25, 2013).

Department of Homeland Security. (2013). Energy sector: critical infrastructure. Available June 2, 2013, at http://www.dhs.gov/energy-sector

McMillan, R. (2008). CIA says hackers have cut power grid. *PC World*, January 19. Available June 2, 2013, at http://www.pcworld.com/article/141564/article.html

Meserve, J. (2007). Staged cyber attack reveals vulnerability in power grid. CNN.com, September 26. Available June 2, 2013, at http://www.cnn.com/2007/US/09/26/power.at.risk/index.html

National Infrastructure Protection Plan (NIPP). (2011). National Infrastructure Protection Plan—Energy Sector. Available June 2, 2013, at http://www.dhs.gov/xlibrary/assets/nppd/nppd-ip-energy-sector-snapshot-2011.pdf

15 Contemporary Safety Management

The new millennium brought a paradigm shift in the United States that completely changed the condition and understanding of the world and most notably our personal safety and security. The United States, with the Cold War over, domestic terrorism at an all-time low, and a thriving economy, achieved a feeling of accomplishment and contentment. Al Qaeda and other terrorist groups had attacked U.S. people and property overseas: the *U.S.S. Cole* in Yemen, soldiers in Somalia, and U.S. embassy bombings in Tanzania and Kenya; however, because of the perception that that was "over there," we felt safe at home in the continental United States. On September 11, 2001, that changed. The attacks on the World Trade Center in New York and the Pentagon in Washington, DC, signaled to the American people that they were not safe. The sense of safety and security that we had long maintained in the United States was gone.

In the United States, many things happened after 9/11 regarding safety and security. There was a different perception of Americans toward the world, and there was a public acceptance, if not demand, for increased security in the homeland. In the years following 9/11, the awareness of threats against our communities required us to be proactive in identifying threats. The PATRIOT Act and the reorganization of government under the Department of Homeland Security (DHS) were answers to that concern. Facility security, awareness programs, and intelligence gathering and sharing are common components of efforts to secure the homeland.

Under the authority of Homeland Security Presidential Directive 7 (HSPD-7), eighteen critical infrastructure sectors were established. These sectors are recognized as areas crucial for the security of the country, and each sector is managed by a sector-specific agency (SSA) that provides sector-level performance feedback to the DHS. In accordance with the National Infrastructure Protection Plan (NIPP), each SSA is responsible for developing and implementing a sector-specific plan (SSP) (Department of Homeland Security, 2007).

U.S. national security was divided into eighteen critical sectors:

1. Food and agriculture
2. Banking and finance
3. Dams
4. Chemical
5. Communications
6. Commercial facilities
7. Critical manufacturing
8. Defense industry base

113

9. Emergency services
10. Energy
11. Government facilities
12. Healthcare and public health
13. Information technology
14. National monuments and icons
15. Nuclear reactors, materials, and waste
16. Postal and shipping
17. Transportation
18. Water

The events of 9/11 brought the recognition that the United States did very little sharing of information in regard to risks and threats, in either public or private organizations. The DHS recognized that each sector already has considerable data available to support a wide range of consequence, risk, and vulnerability assessments. So part of the mission of the DHS is to facilitate the sharing of data that are collected and used by owners, operators, trade associations, and a variety of industry organizations.

Likewise, the DHS collects a wide variety of information, principally through the authorities of various federal agencies and at the state and local levels. The DHS established communication links between federal, state, and local government representatives and industry. During times of increased security posture or emergency situations, the best information sources are the trusted relationships between government and industry.

GOVERNMENT PROGRAMS

INFORMATION SHARING

In a joint effort, the DHS has partnered with the sector SCCs to develop the Homeland Security Information Network (HSIN), an Internet-based communications system that enhances reporting and information sharing and allows industry participants to communicate securely with each other, with other industry sectors, and with government agencies. HSIN has agreements with both public and private sector leadership to share information via a restricted-access network for key personnel to exchange information with the department during emergency situations. The site provides threat awareness and relevant security analyses and presentations (Homeland Security Information Network, 2007).

FOCUSED PROGRAMS

The DHS also facilitated the development of endurance and protection programs with both public and private sector organizations in partnership with the DHS and other appropriate federal agencies. Programs will draw from effective practice already in use by industry and from national laboratory efforts; specific programs will be designed to account for the significant interdependencies between them (Presti, 2012).

Establishing roles and responsibilities for the implementation of new resiliency and protective measures and programs will present both a challenge and an opportunity. For example, in light of increasing cyber threats, the DHS calls on both the public and private sector to work closer together to arrive at a conclusion that would be both equitable and effective. Such an agreement would likely include real-time information sharing and the development of best practices to protect critical infrastructure. (Presti, 2012)

SHARED GOALS FOR SECTOR SECURITY

The DHS supports sector-specific goals for safety and security. Each of the eighteen sectors of critical infrastructure is led by one of various groups, if not a federal agency designated to coordinate goals to enhance security in that area. Goals often relate to response-planning exercises and also to facilitate loss plans that incorporate federal, state, and local law enforcement. The DHS also supports goals to further enhance communication and coordination across the individual sectors, as well as through all sectors.

1. Food and agriculture are overseen by the U.S. Department of Agriculture and the Department of Health and Human Services' Food and Drug Administration.
2. Banking and finance are overseen by the U.S. Department of the Treasury.
3. Dams are overseen by state dam safety offices.
4. Chemicals are overseen as follows: since the majority of chemical sector facilities are privately owned, they require direct relationships with the DHS.
5. Communications are overseen as follows: as with most the private sector, owners and operators of the majority of the communications infrastructures are the primary entities responsible for protecting sector infrastructure and assets.
6. Commercial facilities are overseen as follows: the majority of the facilities in this sector are privately owned and operated, with minimal interaction with the federal government and other regulatory entities.
7. Critical manufacturing is overseen as follows: the majority of the organizations in this sector are privately owned and operated, with minimal interaction with the federal government and other regulatory entities.
8. The defense industry base is overseen by the U.S. Department of Defense.
9. Emergency services have dependencies and interdependencies with multiple critical infrastructure sectors and the National Response Framework's Emergency Support Functions, which supply elements for both the operation and protection of Emergency Services Sector (ESS) assets.
10. Energy is overseen by the U.S. Department of Energy.
11. Government facilities are overseen by the Federal Protective Service and the U.S. Department of Education for the education subsector.

12. Healthcare and public health are overseen by the U.S. Department of Health and Human Services.
13. Information technology is overseen as follows: the majority of information technology assets are privately owned, and thus require direct relationships with the DHS.
14. National monuments and icons are overseen by the U.S. Department of the Interior.
15. Nuclear reactors, materials, and waste are overseen by the Nuclear Regulatory Commission.
16. Postal and shipping are overseen by the Postal and Shipping Government Coordinating Council.
17. Transportation is overseen by the Transportation Security Administration.
18. Water is overseen by the Environmental Protection Agency.

INFORMATION SHARING AND COMMUNICATION

DHS has improved information sharing. While information sharing still may not be optimally utilized, a security manager can get a better understanding of the risk environments now than prior to 9/11. Both the public and private sectors need credible, timely, actionable information to ensure that appropriate security investments, programs, and decisions are made to protect organizational assets. DHS has attempted to build a system of trust in both public and private sector security partners regarding shared information.

The DHS has provided many resources in assisting safety managers to better deal with risk and subsequent loss. The DHS has provided new methods or better explained existing methods that are acceptable to all stakeholders for collecting, protecting, and, as necessary, sharing sensitive data on the vulnerabilities of assets and the protective programs to address them. The private sector will be understandably cautious in providing the information needed for vulnerability assessments and disclosing the results of assessments it has conducted, and it may be equally cautious about providing specifics on ongoing and planned protective programs (Department of Homeland Security, 2007).

CONCLUSION

While the new millennium brought a paradigm shift in the United States affecting our own perceptions of safety and security, the security manager is now in a position to have a better understanding of the risks, threats, vulnerabilities, and criticalities of assets than ever before. For safety managers, there is nothing more practical than a methodical process of thinking. For managers, good thinking pays off while poor thinking causes problems that waste resources and time. Critical thinking allows safety managers to envision their duties in a logical process while focusing on making decisions and solving problems. It has been the intent of this chapter to help the security manager realize the support resources that are available for securing organizational assets in the post-9/11 paradigm.

EXERCISE

This chapter discusses the various sectors of critical infrastructure in the post 9/11 world. An underlying premise relates to infrastructure interdependencies and risk considerations involving cross-sector analysis. In the event of loss sustained by one sector, how could the infrastructure linkages across sectors adversely affect the performance of other infrastructures?

REFERENCES

Department of Homeland Security. (2007). Critical infrastructure sectors. Available May 24, 2013, at http://www.dhs.gov/critical-infrastructure-sectors

Homeland Security Information Network. (2007). [Home page]. Available May 24, 2013, at http://www.dhs.gov/homeland-security-information-network

Presti, K. (2012). DHS secretary calls for public, private partnership to protect critical infrastructure. CRN. October 31. Available May 24, 2013, at http://www.crn.com/news/security/240012631/dhs-secretary-calls-for-public-private-partnership-to-protect-critical-infrastructure.htm

Appendix A: OSHA Recordkeeping

OCCUPATIONAL SAFETY AND HEALTH ADMINISTRATION (OSHA), THE REGULATION AND RELATED INTERPRETATIONS FOR RECORDING AND REPORTING OCCUPATIONAL INJURIES AND ILLNESSES, RECORDKEEPING HANDBOOK, OSHA 3245-01R (U.S. DEPARTMENT OF LABOR, DIRECTORATE OF EVALUATION AND ANALYSIS, OFFICE OF STATISTICAL ANALYSIS, 2005)

Employers are responsible for providing a safe and healthful workplace for their employees. OSHA's role is to assure the safety and health of America's workers by setting and enforcing standards; providing training, outreach, and education; establishing partnerships; and encouraging continual improvement in workplace safety and health. For more information, visit www.osha.gov.

This handbook provides a general overview of a particular topic related to OSHA regulation. It does not alter or determine compliance responsibilities in OSHA standards or the Occupational Safety and Health Act of 1970. Because interpretations and enforcement policy may change over time, you should consult current OSHA administrative interpretations and decisions by the Occupational Safety and Health Review Commission and the Courts for additional guidance on OSHA compliance requirements.

This information is available to sensory impaired individuals upon request. Voice phone: (202) 693-1999; teletypewriter (TTY) number: (877) 889-5627.

INTRODUCTION

This OSHA Web-based Recordkeeping Handbook is a compendium of existing agency-approved recordkeeping materials, including the regulatory text from the 2001 final rule on Occupational Injury and Illness Recording and Reporting Requirements ("the Recordkeeping rule") and relevant explanatory excerpts from the preamble to the rule; chapter 5 of the agency's Recordkeeping Policies and Procedures Manual; Frequently Asked Questions (FAQs); and OSHA letters of interpretation. This Web-based handbook is intended to be a resource for businesses of all sizes, as well as OSHA's compliance safety and health officers, compliance assistance specialists, and OSHA State plans. The information in the handbook is accessible by means of a user-friendly search engine that relies on simple point-and-click technology. The handbook is designed to answer recordkeeping questions raised by employers, employees, and members of the OSHA family who are familiar with the basic requirements of the rule but wish to obtain additional information on specific recordkeeping issues. Users will also find the handbook useful as a research and training tool. The handbook can be accessed at www.osha.gov/recordkeeping/handbook/index.html, which directs users to the file. Because the handbook is Web based, it will be possible to update letters of interpretation and add FAQs to the file quickly as new questions about the rule are raised.

PREFACE

The Occupational Safety and Health Act of 1970 (OSH Act) requires covered employers to prepare and maintain records of occupational injuries and illnesses. The Occupational Safety and Health Administration (OSHA) in the U.S. Department of Labor is responsible for administering the recordkeeping system established by the Act. The OSH Act and recordkeeping regulations in Code of Federal Regulations 29 CFR §1904 and §1952 provide specific recording and reporting requirements which comprise the framework for the nationwide occupational safety and health recording system.

Under this system, it is essential that data recorded by employers be uniform and accurate to assure the consistency and validity of the statistical data which are used by OSHA for many purposes, including inspection targeting, performance measurement under the Government Performance and Results Act (GPRA), standards development, resource allocation, Voluntary Protection Program (VPP) eligibility, and "low-hazard" industry exemptions. The data also aid employers, employees, and compliance officers in analyzing the safety and health environment at the employer's establishment and is the source of information for the OSHA Data Initiative (ODI) and the Bureau of Labor Statistics' (BLS) Annual Survey.

In January 2001, OSHA issued a final rule revising the §1904 and §1952 Occupational Injury and Illness Recording and Reporting Requirements (Recordkeeping) regulations, the first revision since 1978. The goals of the revision were to simplify the system, clarify ongoing concepts, produce more useful information, and better utilize modern technology. The new regulation took effect on January 1, 2002. As part of OSHA's extended outreach efforts, the agency also produced Recordkeeping Policies and Procedures Manual (CPL 2-0.135, revised December 30, 2004), which contained, along with other related information, a variety of Frequently Asked Questions. In addition, in 2002, a detailed Injury and Illness Recordkeeping website was established containing links to helpful resources related to Recordkeeping, including training presentations, applicable Federal Register notices, and OSHA's recordkeeping-related Letters of Interpretation (see http://www.osha.gov/recordkeeping/index.html).

This publication brings together relevant information from the Recordkeeping rule, the policies and procedures manual, and the website. It is organized by regulatory section and contains the specific final regulatory language, selected excerpts from the relevant OSHA decision analysis contained in the preamble to the final rule, along with recordkeeping-related Frequently Asked Questions and OSHA's enforcement guidance presented in the agency's Letters of Interpretation. The user will find this information useful in understanding the Recordkeeping requirements and will be able to easily locate a variety of specific and necessary information pertaining to each section of the rule.

The information included here deals only with the requirements of the Occupational Safety and Health Act of 1970 and Parts 1904 and 1952 of Title 29, Code of Federal Regulations, for recording and reporting occupational injuries and illnesses. Some employers may be subject to additional recordkeeping and reporting requirements not covered in this document. Many specific OSHA standards and regulations have additional requirements for the maintenance and retention of records for medical surveillance, exposure monitoring, inspections, and other activities and incidents relevant to occupational safety and health, and for the reporting of certain information to employees and to OSHA. For information on these requirements, which are not covered in this publication, employers should refer directly to the OSHA standards or regulations, consult OSHA's website for additional information (www.osha.gov), or contact their OSHA regional office or participating State agency.

For recordkeeping and reporting questions not covered in this publication, employers may contact their OSHA regional office or the participating State agency serving their jurisdiction.

REGULATION: SECTION **1904.0**

SUBPART A—PURPOSE **(66 FR 6122, JAN. 19, 2001)**

The purpose of this rule (Part 1904) is to require employers to record and report work-related fatalities, injuries, and illnesses.

Note to Section 1904.0: Recording or reporting a work-related injury, illness, or fatality does not mean that the employer or employee was at fault, that an OSHA rule has been violated, or that the employee is eligible for workers' compensation or other benefits.

Preamble Discussion: Section 1904.0 (66 FR 5933-5935, Jan 19, 2001)

The following are selected excerpts from the preamble to the Occupational Injury and Illness Recording and Reporting Requirements, or the Recordkeeping rule (66 FR 5916, 29 CFR Parts 1904 and 1952). These excerpts represent some of the key discussions related to the final rule (66 FR 6122, 29 CFR Parts 1904 and 1952).

Subpart A. Purpose

The Purpose section of the final rule explains why OSHA is promulgating this rule. The Purpose section contains no regulatory requirements and is intended merely to provide information. A Note to this section informs employers and employees that recording a case on the OSHA recordkeeping forms does not indicate either that the employer or the employee was at fault in the incident or that an OSHA rule has been violated. Recording an injury or illness on the Log also does not, in and of itself, indicate that the case qualifies for workers' compensation or other benefits. Although any specific work-related injury or illness may involve some or all of these factors, the record made of that injury or illness on the OSHA recordkeeping forms only shows three things: (1) that an injury or illness has occurred; (2) that the employer has determined that the case is work-related (using OSHA's definition of that term); and (3) that the case is non-minor, i.e., that it meets one or more of the OSHA injury and illness recording criteria.

Many cases that are recorded in the OSHA system are also compensable under the State workers' compensation system, but many others are not. However, the two systems have different purposes and scopes. The OSHA recordkeeping system is intended to collect, compile, and analyze uniform and consistent nationwide data on occupational injuries and illnesses. The workers' compensation system, in contrast, is not designed primarily to generate and collect data but is intended primarily to provide medical coverage and compensation for workers who are killed, injured, or made ill at work, and varies in coverage from one State to another.

As a result of these differences between the two systems, recording a case does not mean that the case is compensable, or vice versa. When an injury or illness occurs to an employee, the employer must independently analyze the case in light of both the OSHA recording criteria and the requirements of the State workers' compensation system to determine whether the case is recordable or compensable, or both.

Frequently Asked Questions: Section 1904.0
(OSHA Instruction, CPL 2-00.135, Chap. 5)—Purpose

*Question 0-1. Why are employers required to keep records
of work-related injuries and illnesses?*

The OSH Act of 1970 requires the Secretary of Labor to produce regulations that require employers to keep records of occupational deaths, injuries, and illnesses. The records are used for several purposes.

Injury and illness statistics are used by OSHA. OSHA collects data through the OSHA Data Initiative (ODI) to help direct its programs and measure its own performance. Inspectors also use the data during inspections to help direct their efforts to the hazards that are hurting workers.

The records are also used by employers and employees to implement safety and health programs at individual workplaces. Analysis of the data is a widely recognized method for discovering workplace safety and health problems and for tracking progress in solving those problems.

The records provide the base data for the Bureau of Labor Statistics annual survey of occupational injuries and illnesses, the nation's primary source of occupational injury and illness data.

*Question 0-2. What is the effect of workers' compensation reports
on the OSHA records?*

The purpose section of the rule includes a note to make it clear that recording an injury or illness neither affects a person's entitlement to workers' compensation nor proves a violation of an OSHA rule. The rules for compensability under workers' compensation differ from state to state and do not have any effect on whether or not a case needs to be recorded on the OSHA 300 Log. Many cases will be OSHA recordable and compensable under workers' compensation. However, some cases will be compensable but not OSHA recordable, and some cases will be OSHA recordable but not compensable under workers' compensation.

Letters of Interpretation: Section 1904.0—Purpose

This section will be developed as letters of interpretation become available.

SECTION 1904.1

PARTIAL EXEMPTION FOR EMPLOYERS WITH 10 OR FEWER EMPLOYEES

(66 FR 6122, JAN. 19, 2001)

REGULATION: SECTION 1904.1

SUBPART B—SCOPE (66 FR 6122, JAN. 19, 2001)

Note to Subpart B: All employers covered by the Occupational Safety and Health Act (OSH Act) are covered by these Part 1904 regulations. However, most employers do not have to keep OSHA injury and illness records unless OSHA or the Bureau of Labor Statistics (BLS) informs them in writing that they must keep records. For example, employers with 10 or fewer employees and business establishments in certain industry classifications are partially exempt from keeping OSHA injury and illness records.

Section 1904.1 Partial Exemption for Employers with 10 or Fewer Employees

(a) Basic Requirement.

(1) If your company had ten (10) or fewer employees at all times during the last calendar year, you do not need to keep OSHA injury and illness records unless OSHA or the BLS informs you in writing that you must keep records under Section 1904.41 or Section 1904.42. However, as required by Section 1904.39, all employers covered by the OSH Act must report to OSHA any workplace incident that results in a fatality or the hospitalization of three or more employees.

(2) If your company had more than ten (10) employees at any time during the last calendar year, you must keep OSHA injury and illness records unless your establishment is classified as a partially exempt industry under Section 1904.2.

(b) Implementation.

(1) Is the partial exemption for size based on the size of my entire company or on the size of an individual business establishment?

The partial exemption for size is based on the number of employees in the entire company.

(2) How do I determine the size of my company to find out if I qualify for the partial exemption for size?

To determine if you are exempt because of size, you need to determine your company's peak employment during the last calendar year. If you had no more than 10 employees at any time in the last calendar year, your company qualifies for the partial exemption for size.

Preamble Discussion: Section 1904.1 (66 FR 5935-5939, Jan. 19, 2001)

The following are selected excerpts from the preamble to the Occupational Injury and Illness Recording and Reporting Requirements, the Recordkeeping rule (66 FR 5916, 29 CFR Parts 1904 and 1952). These excerpts represent some of the key discussions related to the final rule (66 FR 6122, 29 CFR Parts 1904 and 1952).

Section 1904.1 Partial Exemption for Employers with 10 or Fewer Employees

The Size-Based Exemption in the Former Rule

The original OSHA injury and illness recording and reporting rule issued in July 1971 required all employers covered by the OSH Act to maintain injury and illness records. In October 1972, an exemption from most of the recordkeeping requirements was put in place for employers with seven or fewer employees. In 1977, OSHA amended the rule to exempt employers with 10 or fewer employees, and that exemption has continued in effect to this day...

The Size-Based Exemption in the Final Rule

...Under the final rule (and the former rule), an employer in any industry who employed no more than 10 employees at any time during the preceding calendar year is not required to maintain OSHA records of occupational illnesses and injuries during the current year unless requested to do so in writing by OSHA (under Section 1904.41) or the BLS (under Section 1904.42). If an employer employed 11 or more people at a given time during the year, how-ever, that employer is not eligible for the size-based partial exemption...

... [U]nder the 10 or fewer employee partial exemption threshold, more than 80% of employers in OSHA's jurisdiction are exempted from routinely keeping records...

...[T]he final rule clarifies that the 10 or fewer size exemption is applicable only if the employer had fewer than 11 employees at all times during the previous calendar year. Thus, if an employer employs 11 or more people at any given time during that year, the employer is not eligible for the small employer exemption in the following year. This total includes all workers employed by the business. All individuals who are "employees" under the OSH Act are counted in the total; the count includes all full time, part time, temporary, and seasonal employees. For businesses that are sole proprietorships or partnerships, the owners and partners would not be considered employees and would not be counted. Similarly, for family farms, family members are not counted as employees. However, in a corporation, corporate officers who receive payment for their services are considered employees (see Section 1904.31, "Covered Employees").

Consistent with the former rule, the final rule applies the size exemption based on the total number of employees in the firm, rather than the number of employees at any particular location or establishment...because the resources available in a given business depend on the size of the firm as a whole, not on the size of individual establishments owned by the firm. In addition, the analysis of injury records should be of value to the firm as a whole, regardless of the size of individual establishments. Further, an exemption based on individual establishments would be difficult to administer, especially in cases where an individual employee, such as a maintenance worker, regularly reports to work at several establishments.

SECTION 1904.2

PARTIAL EXEMPTION FOR ESTABLISHMENTS IN CERTAIN INDUSTRIES

(66 FR 6122, JAN. 19, 2001)

REGULATION: SECTION 1904.2

SUBPART B—SCOPE (66 FR 6122, JAN. 19, 2001)

SECTION 1904.2 PARTIAL EXEMPTION FOR ESTABLISHMENTS IN CERTAIN INDUSTRIES

(a) Basic requirement.

(1) If your business establishment is classified in a specific low hazard retail, service, finance, insurance or real estate industry listed in Appendix A to this Subpart B, you do not need to keep OSHA injury and illness records unless the government asks you to keep the records under Section 1904.41 or Section 1904.42. However, all employers must report to OSHA any workplace incident that results in a fatality or the hospitalization of three or more employees (see Section 1904.39).

(2) If one or more of your company's establishments are classified in a non-exempt industry, you must keep OSHA injury and illness records for all of such establishments unless your company is partially exempted because of size under Section 1904.1.

(b) Implementation.

(1) Does the partial industry classification exemption apply only to business establishments in the retail, services, finance, insurance or real estate industries (SICs 52-89)?

Yes, business establishments classified in agriculture; mining; construction; manufacturing; transportation; communication; electric, gas, and sanitary services; or wholesale trade are not eligible for the partial industry classification exemption.

(2) Is the partial industry classification exemption based on the industry classification of my entire company or on the classification of individual business establishments operated by my company?

The partial industry classification exemption applies to individual business establishments. If a company has several business establishments engaged in different classes of business activities, some of the company's establishments may be required to keep records, while others may be exempt.

(3) How do I determine the Standard Industrial Classification code for my company or for individual establishments?

You determine your Standard Industrial Classification (SIC) code by using the Standard Industrial Classification Manual, Executive Office of the President, Office of Management and Budget. You may contact your nearest OSHA office or State agency for help in determining your SIC.

For additional information see http://www.osha.gov/recordkeeping/handbook/index.html

Appendix B: OSHA—Recording and Reporting Occupational Injuries and Illness

29 CFR: 1904: RECORDING AND REPORTING OCCUPATIONAL INJURIES AND ILLNESS; REPORTING FATALITY, INJURY, AND ILLNESS INFORMATION TO THE GOVERNMENT

1904.39: REPORTING FATALITIES AND MULTIPLE HOSPITALIZATION INCIDENTS TO OSHA.

1904.39(A)

Basic requirement. Within eight (8) hours after the death of any employee from a work-related incident or the in-patient hospitalization of three or more employees as a result of a work-related incident, you must orally report the fatality/multiple hospitalization by telephone or in person to the Area Office of the Occupational Safety and Health Administration (OSHA), U.S. Department of Labor, that is nearest to the site of the incident. You may also use the OSHA toll-free central telephone number, 800-321-OSHA (800-321-6742).

1904.39(b)

Implementation.

1904.39(b)(1)

If the Area Office is closed, may I report the incident by leaving a message on OSHA's answering machine, faxing the area office, or sending an e-mail? No, if you can't talk to a person at the Area Office, you must report the fatality or multiple hospitalization incident using the 800 number.

1904.39(b)(2)

What information do I need to give to OSHA about the incident? You must give OSHA the following information for each fatality or multiple hospitalization incident:

1904.39(b)(2)(i)

The establishment name;

1904.39(b)(2)(ii)
The location of the incident;

1904.39(b)(2)(iii)
The time of the incident;

1904.39(b)(2)(iv)
The number of fatalities or hospitalized employees;

1904.39(b)(2)(v)
The names of any injured employees;

1904.39(b)(2)(vi)
Your contact person and his or her phone number; and

1904.39(b)(2)(vii)
A brief description of the incident.

1904.39(b)(3)
Do I have to report every fatality or multiple hospitalization incident resulting from a motor vehicle accident? No, you do not have to report all of these incidents. If the motor vehicle accident occurs on a public street or highway, and does not occur in a construction work zone, you do not have to report the incident to OSHA. However, these injuries must be recorded on your OSHA injury and illness records, if you are required to keep such records.

1904.39(B)(4)

Do I have to report a fatality or multiple hospitalization incident that occurs on a commercial or public transportation system? No, you do not have to call OSHA to report a fatality or multiple hospitalization incident if it involves a commercial airplane, train, subway, or bus accident. However, these injuries must be recorded on your OSHA injury and illness records, if you are required to keep such records.

1904.39(B)(5)

Do I have to report a fatality caused by a heart attack at work? Yes, your local OSHA Area Office director will decide whether to investigate the incident, depending on the circumstances of the heart attack.

1904.39(B)(6)

Do I have to report a fatality or hospitalization that occurs long after the incident? No, you must only report each fatality or multiple hospitalization incident that occurs within thirty (30) days of an incident.

1904.39(B)(7)

What if I don't learn about an incident right away? If you do not learn of a reportable incident at the time it occurs and the incident would otherwise be reportable under paragraphs (a) and (b) of this section, you must make the report within eight (8) hours of the time the incident is reported to you or to any of your agent(s) or employee(s).

[66 FR 6133, Jan. 19, 2001]

For additional information, see http://www.osha.gov/pls/oshaweb/owadisp.show_document?p_table=STANDARDS&p_id = 12783

Appendix C: Chemical Facility Anti-Terrorism Standards[*]

---

[*] Chemical Facility Anti-Terrorism Standards, Final Rule, 72 Fed. Reg. 17687–17745 (Apr. 9, 2007).

Chemicals of Interest	Synonym	Theft Minimum Concentration (%)	Theft Screening Threshold Quantities (in Pounds unless Otherwise Noted)	Sabotage Minimum Concentration (%)	Sabotage Screening Threshold Quantities (in Pounds)	Release –Toxic	Release –Flammables	Release –Explosives	Theft–Chemical Weapons (CWs) and Chemical Weapons Precursors (CWPs)	Theft–Weapons of Mass Effect (WMEs)	Theft–Explosives (EXPs) or Improvised Explosive Device Precursors (IEDPs)	Sabotage or Contamination
Acetaldehyde							X					
Acetone cyanohydrin, stabilized				ACG	APA							X
Acetyl bromide				ACG	APA							X
Acetyl chloride				ACG	APA							X
Acetyl iodide				ACG	APA							X
Acetylene	Ethyne						X					
Acrolein	2-Propenal; acrylaldehyde					X						
Acrylonitrile	2-Propenenitrile						X					
Acrylyl chloride	2-Propenoyl chloride						X					
Allyl alcohol	2-Propen-1-ol					X						
Allylamine	2-Propen-1-amine						X					
Allyltrichlorosilane, stabilized				ACG	APA							X
Aluminum (powder)		ACG	100								X	
Aluminum bromide, anhydrous				ACG	APA							X
Aluminum chloride, anhydrous				ACG	APA							X
Aluminum phosphide				ACG	APA							X
Ammonia (anhydrous)				ACG	APA	X						

Chemical	1	2	3	4	5	Min Conc	STQ	Code	Code	6	7
Ammonia (concentration of 20% or greater)		X			X	ACG	400				X
Ammonium nitrate (with more than 0.2% combustible substances, including any organic substance calculated as carbon, to the exclusion of any other added substance)		X									X
Ammonium nitrate, solid (nitrogen concentration of 23% nitrogen or greater)		X				33	2000				X
Ammonium perchlorate		X			X	ACG	400				X
Ammonium picrate	X	X			X	ACG	400				X
Amyltrichlorosilane								ACG	APA		X
Antimony pentafluoride	X			X				ACG	APA		X
Arsenic trichloride (Arsenous trichloride)				X		30	2.2			X	X
Arsine		X			X	0.67	15			X	X
Barium azide					X	ACG	400				
1,4-Bis(2-chloroethylthio)-nbutane					X	CUM 100g					
Bis(2-chloroethylthio)methane					X	CUM 100g					
Bis(2-chloroethylthiomethyl)ether					X	CUM 100g					
1,5-Bis(2-chloroethylthio)-npentane					X	CUM 100g					
1,3-Bis(2-chloroethylthio)-npropane					X	CUM 100g					
Boron tribromide			X			12.67	45	ACG	APA		X
Boron trichloride (Borane, trichloro)			X			84.7	45				X
Boron trifluoride (Borane, trifluoro)			X			26.87	45				X
Boron trifluoride compound with methyl ether (1:1) (Boron, trifluoro [oxybis] (methane), T-4)	X										X
Bromine											X

Legend: ACG = a commercial grade; APA = a placarded amount.

Continued

Chemicals of Interest	Synonym	Theft Minimum Concentration (%)	Theft Screening Threshold Quantities (in Pounds unless Otherwise Noted)	Sabotage Minimum Concentration (%)	Sabotage Screening Threshold Quantities (in Pounds)	Release–Toxic	Release–Flammables	Release–Explosives	Theft–Chemical Weapons (CWs) and Chemical Weapons Precursors (CWPs)	Theft–Weapons of Mass Effect (WMEs)	Theft–Explosives (EXPs) or Improvised Explosive Device Precursors (IEDPs)	Sabotage or Contamination
Bromine chloride		9.67	45							X		
Bromine pentafluoride				ACG	APA							X
Bromotrifluoroethylene	Ethene, bromotrifluoro-						X					
1,3-Butadiene							X					
Butane							X					
Butene							X					
1-Butene							X					
2-Butene							X					
2-Butene-cis							X					
2-Butene-trans	2-Butene, (E)						X					
Butyltrichlorosilane				ACG	APA							X
Calcium hydrosulfite	Calcium dithionite			ACG	APA							X
Calcium phosphide				ACG	APA							X
Carbon disulfide						X						
Carbon oxysulfide	Carbon oxide sulfide (COS); carbonyl sulfide						X					
Carbonyl fluoride		12	45							X		
Carbonyl sulfide		56.67	500							X		

Chemical	Synonym	Min Conc	STQ							
Chlorine		9.77	500	ACG	APA	X	X		X	X
Chlorine dioxide	Chlorine oxide, (ClO2)					X	X			
Chlorine monoxide	Chlorine oxide									
Chlorine pentafluoride		4.07	15						X	
Chlorine trifluoride		9.97	45						X	X
Chloroacetyl chloride				ACG	APA					
2-Chloroethylchloromethylsulfide			CUM 100g				X			
Chloroform	Methane, trichloro-					X				
Chloromethyl methyl ether	Methane, chloromethoxy-					X				
1-Chloropropylene	1-Propene, 1-chloro-						X			
2-Chloropropylene	1-Propene, 2-chloro-						X			
Chlorosarin	o-Isopropyl methylphosphonochloridate		CUM 100g					X		
Chlorosoman	o-Pinacolyl methylphosphonochloridate		CUM 100g					X		
Chlorosulfonic acid				ACG	APA					X
Chromium oxychloride				ACG	APA					X
Crotonaldehyde	2-Butenal						X			
Crotonaldehyde, (E)-	2-Butenal, (E)-		45				X			
Cyanogen	Ethanedinitrile	11.67	45				X	X	X	
Cyanogen chloride		2.67	15			X			X	
Cyclohexanamine						X				
Cyclohexyltrichlorosilane				ACG	APA		X			X
Cyclopropane										
DF	Methyl phosphonyl difluoride		CUM 100g					X		
Diazodinitrophenol		ACG	400					X		X
Diborane		2.67	15			X			X	
Dichlorosilane	Silane, dichloro-	10.47	45				X		X	

Legend: ACG = a commercial grade; APA = a placarded amount.

Continued

Chemicals of Interest	Synonym	Theft — Minimum Concentration (%)	Theft — Screening Threshold Quantities (in Pounds unless Otherwise Noted)	Sabotage — Minimum Concentration (%)	Sabotage — Screening Threshold Quantities (in Pounds)	Release–Toxic	Release–Flammables	Release–Explosives	Theft–Chemical Weapons (CWs) and Chemical Weapons Precursors (CWPs)	Theft–Weapons of Mass Effect (WMEs)	Theft–Explosives (EXPs) or Improvised Explosive Device Precursors (IEDPs)	Sabotage or Contamination
N,N-(2-diethylamino)ethanethiol		30	2.2						X			
Diethyldichlorosilane				ACG	APA							X
Diethyleneglycol dinitrate		ACG	400					X			X	
Diethyl methylphosphonite		30	2.2						X			
N,N-diethyl phosphoramidic dichloride		30	2.2						X			
N,N-(2-diisopropylamino)ethanethiol		30	2.2						X			
N,N-diisopropyl-(beta)-aminoethane thiol												
Difluoroethane	Ethane, 1,1-difluoro-						X					
N,N-diisopropyl phosphoramidic dichloride		30	2.2						X			
1,1-Dimethylhydrazine	Hydrazine, 1,1-dimethyl-						X					
Dimethylamine	Methanamine, N-methyl-						X					
N,N-(2-dimethylamino)ethanethiol		30	2.2				X		X			
Dimethyldichlorosilane	Silane, dichlorodimethyl-	30	2.2	ACG	APA							X
N,N-dimethyl phosphoramidic dichloride		30	2.2						X			
Dimethylphosphoramidodichloridate												

Chemical	Synonym	Min Conc	STQ								
2,2-Dimethylpropane	Propane, 2,2-dimethyl-	ACG	400				X				
Dingu	Dinitroglycoluril	3.8	15						X		X
Dinitrogen tetroxide		ACG	400				X			X	
Dinitrophenol		ACG	400				X		X	X	X
Dinitroresorcinol		ACG	APA								X
Diphenyldichlorosilane		ACG	400				X				X
Dipicryl sulfide		ACG	400					X			
N,N-(2-dipropylamino)ethanethiol		30	2.2					X			
N,N-dipropyl phosphoramidic dichloride		30	2.2					X			
Dodecyltrichlorosilane		ACG	APA								X
Epichlorohydrin	Oxirane, (chloromethyl)-					X					
Ethane							X				
Ethyl acetylene	1-Butyne						X				
Ethyl chloride	Ethane, chloro-						X				
Ethyl ether	Ethane, 1,1-oxybis-						X				
Ethyl mercaptan	Ethanethiol						X				
Ethyl nitrite	Nitrous acid, ethyl ester						X				
Ethyl phosphonyl difluoride		CUM	100g				X				
Ethylamine	Ethanamine						X	X			
Ethyldiethanolamine		80	220					X			
Ethylene	Ethene						X				
Ethylene oxide	Oxirane					X	X				
Ethylenediamine	1,2-Ethanediamine					X					
Ethyleneimine	Aziridine						X				
Ethylphosphonothioic dichloride		30	2.2					X			
Ethyltrichlorosilane		ACG	APA								X
Fluorine		6.17	15							X	X
Fluorosulfonic acid		ACG	APA								X

Legend: ACG = a commercial grade; APA = a placarded amount.

Continued

Chemicals of Interest	Synonym	Theft — Minimum Concentration (%)	Theft — Screening Threshold Quantities (in Pounds unless Otherwise Noted)	Sabotage — Minimum Concentration (%)	Sabotage — Screening Threshold Quantities (in Pounds)	Release–Toxic	Release–Flammables	Release–Explosives	Theft–Chemical Weapons (CWs) and Chemical Weapons Precursors (CWPs)	Theft–Weapons of Mass Effect (WMEs)	Theft–Explosives (EXPs) or Improvised Explosive Device Precursors (IEDPs)	Sabotage or Contamination
Formaldehyde (solution)						X						
Furan							X					
Germanium tetrafluoride		2.11	15							X		
Guanyl nitrosaminoguanylidene hydrazine		ACG	400								X	
Hexaethyl tetraphosphate and compressed gas mixtures		33.37	500							X		
Hexafluoroacetone		15.67	45							X		
Hexanitrostilbene		ACG	400					X			X	
Hexolite	Hexotol	ACG	400					X			X	
Hexyltrichlorosilane				ACG	APA							X
HMX	Cyclotetramethylenetetranitramine	ACG	400					X			X	
HN1 (nitrogen mustard-1)	Bis(2-chloroethyl)ethylamine		CUM 100g						X			
HN2 (nitrogen mustard-2)	Bis(2-chloroethyl)methylamine		CUM 100g						X			
HN3 (nitrogen mustard-3)	Tris(2-chloroethyl)amine		CUM 100g						X			
Hydrazine							X					
Hydrochloric acid (concentration of 37% or greater)						X						

Chemical	Synonym / CAS	Min Conc	STQ	1	2	3	4	5	6	7	8
Hydrocyanic acid				X	X						
Hydrofluoric acid (concentration of 50% or greater)				X	X						
Hydrogen						X					
Hydrogen bromide (anhydrous)		95.33	500			X			X		
Hydrogen chloride (anhydrous)		ACG	500	X					X		
Hydrogen cyanide	Hydrocyanic acid	4.67	15	X					X		
Hydrogen fluoride (anhydrous)		42.53	45	X					X		
Hydrogen peroxide (concentration of at least 35%)		35	400							X	
Hydrogen selenide		0.07	15			X			X		
Hydrogen sulfide		23.73	45	X					X		X
Iodine pentafluoride		ACG	APA								X
Iron, pentacarbonyl-	Iron carbonyl (Fe(CO)5), (TB5-11)-					X	X				
Isobutane	Propane, 2-methyl					X					
Isobutyronitrile	Propanenitrile, 2-methyl-			X		X					
Isopentane	Butane, 2-methyl-			X		X					
Isoprene	1,3-Butadiene, 2-methyl-			X		X					
Isopropyl chloride	Propane, 2-chloro-			X		X					
Isopropyl chloroformate	Carbonochloridic acid, 1-methylethyl ester			X		X					
Isopropylamine	2-Propanamine	30	2.2					X			
Isopropylphosphonothioic dichloride			CUM 100g				X				
Isopropylphosphonyl difluoride			CUM 100g				X				
Lead azide		ACG	400				X			X	
Lead styphnate	Lead trinitroresorcinate	ACG	400				X			X	
Lewisite 1	2-Chlorovinyldichloroarsine		CUM 100g				X				
Lewisite 2	Bis(2-chlorovinyl)chloroarsine		CUM 100g				X				

Legend: ACG = a commercial grade; APA = a placarded amount.

Continued

Chemicals of Interest	Synonym	Theft Minimum Concentration (%)	Theft Screening Threshold Quantities (in Pounds unless Otherwise Noted)	Sabotage Minimum Concentration (%)	Sabotage Screening Threshold Quantities (in Pounds)	Release–Toxic	Release–Flammables	Release–Explosives	Theft–Chemical Weapons (CWs) and Chemical Weapons Precursors (CWPs)	Theft–Weapons of Mass Effect (WMEs)	Theft–Explosives (EXPs) or Improvised Explosive Device Precursors (IEDPs)	Sabotage or Contamination
Lewisite 3	Tris(2-chlorovinyl)arsine		CUM 100g	ACG	APA			X				
Lithium amide												X
Magnesium (powder)		ACG	100								X	
Magnesium diamide				ACG	APA							X
Magnesium phosphide				ACG	APA							X
MDEA	Methyldiethanolamine	80	220						X			
Mercury fulminate		ACG	400					X			X	
Methacrylonitrile	2-Propenenitrile, 2-methyl-					X						
Methane							X					
2-Methyl-1-butene							X					
3-Methyl-1-butene							X					
Methyl chloride	Methane, chloro-						X					
Methyl chloroformate	Carbonochloridic acid, methyl ester						X					
Methyl ether	Methane, oxybis-						X					
Methyl formate	Formic acid, methyl ester						X					
Methyl hydrazine	Hydrazine, methyl-					X						
Methyl isocyanate	Methane, isocyanato-					X						
Methyl mercaptan	Methanethiol	45	500				X					

Chemical	Synonym	Value 1	Value 2	Type 1	Type 2										
Methyl thiocyanate	Thiocyanic acid, methyl ester					X									
Methylamine	Methanamine	20	45				X					X			
Methylchlorosilane				ACG	APA										X
Methyldichlorosilane				ACG	APA										X
Methylphenyldichlorosilane		30	2.2						X						
Methylphosphonothioic dichloride															X
Methyltrichlorosilane	Silane, trichloromethyl-			ACG	APA		X								
Sulfur mustard (mustard gas (H))	Bis(2-chloroethyl)sulfide	CUM 100g						X							
O-mustard (T)	Bis(2-chloroethylthioethyl)ether	CUM 100g						X							
Nickel carbonyl							X								
Nitric acid		68	400			X		X					X		
Nitric oxide	Nitrogen oxide (NO)	3.83	15			X		X					X		
Nitrobenzene		ACG	100										X		
5-Nitrobenzotriazol		ACG	400							X			X		
Nitrocellulose		ACG	400							X			X		
Nitrogen mustard hydrochloride	Bis(2-chloroethyl)methylamine hydrochloride	30	2.2								X				
Nitrogen trioxide		3.83	15									X			
Nitroglycerine		ACG	400							X			X		
Nitromannite	Mannitol hexanitrate, wetted	ACG	400							X			X		
Nitromethane		ACG	400							X			X		
Nitrostarch		ACG	400							X			X		
Nitrosyl chloride		1.17	15									X	X		
Nitrotriazolone		ACG	400							X			X		
Nonyltrichlorosilane				ACG	APA										X
Octadecyltrichlorosilane				ACG	APA										X
Octolite		ACG	400							X			X		
Octonal		ACG	400							X			X		

Continued

Legend: ACG = a commercial grade; APA = a placarded amount.

Chemicals of Interest	Synonym	Theft Minimum Concentration (%)	Theft Screening Threshold Quantities (in Pounds unless Otherwise Noted)	Sabotage Minimum Concentration (%)	Sabotage Screening Threshold Quantities (in Pounds)	Release–Toxic	Release–Flammables	Release–Explosives	Theft–Chemical Weapons (CWs) and Chemical Weapons Precursors (CWPs)	Theft–Weapons of Mass Effect (WMEs)	Theft–Explosives (EXPs) or Improvised Explosive Device Precursors (IEDPs)	Sabotage or Contamination
Oleum (fuming sulfuric acid)	Sulfuric acid, mixture with sulfur trioxide					X						
Oxygen difluoride		0.09	15							X		
1,3-Pentadiene							X					
Pentane							X					
1-Pentene							X					
2-Pentene, (E)-							X					
2-Pentene, (Z)-							X					
Pentolite		ACG	400					X			X	
Peracetic acid	Ethaneperoxic acid						X					
Perchloromethylmercaptan	Methanesulfenyl chloride, trichloro-					X						
Perchloryl fluoride		25.67	45							X		
PETN	Pentaerythritol tetranitrate	ACG	400					X			X	
Phenyltrichlorosilane				ACG	APA							X
Phosgene	Carbonic dichloride; carbonyl dichloride	0.17	15			X				X		
Phosphine		0.67	15				X			X		
Phosphorus		ACG	400								X	

Chemical of Interest	Synonym	Min Conc	STQ	APA	Release–Toxic	Release–Flammable	Release–Explosive	Theft–CW/CWP	Theft–WME	Theft–EXP/IEDP	Sabotage
Phosphorus oxychloride	Phosphoryl chloride	ACG	80	APA	X			X			X
Phosphorus pentabromide		ACG	220	APA							X
Phosphorus pentachloride		ACG		APA							X
Phosphorus pentasulfide		ACG		APA					X		X
Phosphorus trichloride		ACG	3.48 / 45	APA	X						X
Piperidine						X		X			
Potassium chlorate		ACG	400							X	
Potassium cyanide		ACG		APA							X
Potassium nitrate		ACG	400							X	
Potassium perchlorate		ACG	400							X	
Potassium permanganate		ACG	400							X	
Potassium phosphide		ACG		APA							X
1,2-Propadiene						X	X				
Propane						X					
Propionitrile					X						
Propyl chloroformate	Carbonchloridic acid, propylester					X					
Propylene [1-propene]						X					
Propylene oxide	Oxirane, methyl-					X					
Propyleneimine	Aziridine, 2-methyl-				X	X					
Propylphosphonothioic dichloride			30				X	X			
Propylphosphonyl difluoride		CUM 100g	2.2					X			
Propyltrichlorosilane		ACG		APA							X
Propyne	1-Propyne					X					
QL	o-Ethyl-o-2-diisopropylaminoethyl methylphosphonite	CUM 100g						X			
RDX	Cyclotrimethylenetrinitramine	ACG	400				X			X	
RDX and HMX mixtures		ACG	400				X			X	
Selenium hexafluoride		1.67	15						X		

Legend: ACG = a commercial grade; APA = a placarded amount.

Continued

Chemicals of Interest	Synonym	Theft Min. Conc. (%)	Theft Screening Threshold Quantities (in Pounds unless Otherwise Noted)	Sabotage Min. Conc. (%)	Sabotage Screening Threshold Quantities (in Pounds)	Release–Toxic	Release–Flammables	Release–Explosives	Theft–Chemical Weapons (CWs) and Chemical Weapons Precursors (CWPs)	Theft–Weapons of Mass Effect (WMEs)	Theft–Explosives (EXPs) or Improvised Explosive Device Precursors (IEDPs)	Sabotage or Contamination
Sesquimustard	1,2-Bis(2-chloroethylthio)ethane	CUM 100g						X				
Silane							X					
Silicon tetrachloride				ACG	APA							X
Silicon tetrafluoride		1.5	45							X		
Sodium azide		ACG	400								X	
Sodium chlorate		ACG	400								X	
Sodium cyanide				ACG	APA							X
Sodium hydrosulfite	Sodium dithionite			ACG	APA							X
Sodium nitrate		ACG	400								X	
Sodium phosphide				ACG	APA							X
Soman	o-Pinacolyl methylphosphonofluoridate	CUM 100g						X				
Stibine		0.67	15							X		
Strontium phosphide				ACG	APA							X
Sulfur dioxide (anhydrous)		84	500			X				X		
Sulfur tetrafluoride	Sulfur fluoride (SF4), (T-4)-	1.33	15			X				X		
Sulfur trioxide						X						
Sulfuryl chloride				ACG	APA							X

Chemical	Synonym	CUM 100g									
Tabun	o-Ethyl-N,Ndimethylphosphoramido-cyanidate	0.83						X			
Tellurium hexafluoride			15						X		
Tetrafluoroethylene	Ethene, tetrafluoro-					X			X		
Tetramethylsilane	Silane, tetramethyl-					X			X		
Tetranitroaniline		ACG	400				X		X		
Tetranitromethane	Methane, tetranitro-					X					
Tetrazene	Guanyl nitrosaminoguanyltetrazene	ACG	400				X	X	X		
1H-retrazole		ACG	400				X	X	X		
Thiodiglycol	Bis(2-hydroxyethyl)sulfide	30	2.2					X	X		
Thionyl chloride		ACG	APA								X
Titanium tetrachloride	Titanium chloride (TiCl4) (T-4)-	13.33	45	ACG	APA			X			X
TNT	Trinitrotoluene	ACG	400				X		X		
Torpex	Hexotonal	ACG	400				X		X		
Trichlorosilane	Silane, trichloro-	ACG	APA			X					X
Triethanolamine		80	220					X			
Triethanolamine hydrochloride		80	220					X			
Triethyl phosphate		80	220					X			
Trifluoroacetyl chloride		6.93	45						X		
Trifluorochloroethylene	Ethene, chlorotrifluoro-	66.67	500			X			X		
Trimethylamine	Meth anamine, N,N-dimethyl-					X					
Trimethylchlorosilane	Silane, chlorotrimethyl-	ACG	APA			X					X
Trimethyl phosphate		80	220					X			
Trinitroaniline		ACG	400				X		X		
Trinitroanisole		ACG	400				X		X		
Trinitrobenzenesulfonic acid		ACG	400				X		X		
Trinitrobenzoic acid		ACG	400				X		X		

Legend: ACG = a commercial grade; APA = a placarded amount.

Continued

Chemicals of Interest	Synonym	Theft Minimum Concentration (%)	Theft Screening Threshold Quantities (in Pounds unless Otherwise Noted)	Sabotage Minimum Concentration (%)	Sabotage Screening Threshold Quantities (in Pounds)	Release–Toxic	Release–Flammables	Release–Explosives	Theft–Chemical Weapons (CWs) and Chemical Weapons Precursors (CWPs)	Theft–Weapons of Mass Effect (WMEs)	Theft–Explosives (EXPs) or Improvised Explosive Device Precursors (IEDPs)	Sabotage or Contamination
Trinitrochlorobenzene		ACG	400					X			X	
Trinitrofluorenone		ACG	400					X			X	
Trinitro-meta-cresol		ACG	400					X			X	
Trinitronaphthalene		ACG	400					X			X	
Trinitrophenetole		ACG	400					X			X	
Trinitrophenol		ACG	400					X			X	
Trinitroresorcinol		ACG	400					X			X	
Tritonal		ACG	400					X			X	
Tungsten hexafluoride		7.1	45							X		
Vinyl acetate monomer	Acetic acid ethenyl ester						X					
Vinyl acetylene	1-Buten-3-yne						X					
Vinyl chloride	Ethene, chloro-						X					
Vinyl ethyl ether	Ethene, ethoxy-						X					
Vinyl fluoride	Ethene, fluoro-						X					
Vinyl methyl ether	Ethene, methoxy-						X					
Vinylidene chloride	Ethene, 1,1-dichloro-						X					
Vinylidene fluoride	Ethene, 1,1-difluoro-						X					

			ACG	APA						
Vinyltrichlorosilane										X
VX	o-Ethyl-S-2-diisopropylaminoethyl methyl phosphonothiolate	CUM 100g					X			
Zinc hydrosulfite	Zinc dithionite		ACG	APA						X

Legend: ACG = a commercial grade; APA = a placarded amount.

Appendix D: Glossary

accidental hazard: a source of harm or difficulty created by negligence, error, or unintended failure.

accountability: the security goal that generates the requirement for actions of an entity to be traced uniquely to that entity.

adversary: an individual, group, organization, or government that conducts or has the intent to conduct detrimental activities.

asset: a person, structure, facility, information, material, or process that has value.

assurance: confidence that the other four security goals (integrity, availability, confidentiality, and accountability) have been adequately met.

attack method: the manner and means, including the weapon and delivery method, an adversary may use to cause harm to a target.

attack path: the steps that an adversary takes or may take to plan, prepare for, and execute an attack.

availability: the security goal that generates the requirement for protection against intentional or accidental attempts to (1) perform unauthorized deletion of data or (2) otherwise cause a denial of service or data.

capability: the means to accomplish a mission, function, or objective.

confidentiality: the security goal that generates the requirement for protection from intentional or accidental attempts to perform unauthorized data reads.

consequence: the effect of an event, incident, or occurrence.

consequence assessment: the process of identifying or evaluating the potential or actual effects of an event, incident, or occurrence.

countermeasure: the action, measure, or device that reduces an identified risk.

deterrent: a measure that discourages an action or prevents an occurrence by instilling fear, doubt, or anxiety.

due care: a duty to provide for information security to ensure that the type of control, the cost of control, and the deployment of control are appropriate for the system being managed.

economic consequence: the effect of an incident, event, or occurrence on the value of property or on the production, trade, distribution, or use of income, wealth, or commodities.

evaluation: the process of examining, measuring, and/or judging how well an entity, procedure, or action has met or is meeting stated objectives.

function: a service, process, capability, or operation performed by an asset, system, network, or organization.

hazard: a natural or human-made source or cause of harm or difficulty.

human consequence: the effect of an incident, event, or occurrence that results in injury, illness, or loss of life.

implementation: the act of putting a procedure or course of action into effect to support goals or achieve objectives.

incident: an occurrence, caused by either human action or natural phenomena, that may cause harm and may require action.

integrated risk management: the incorporation and coordination of strategy, capability, and governance to enable risk-informed decision making.

integrity: the security goal that generates the requirement for protection against either intentional or accidental attempts to violate data integrity.

intent: a determination to achieve an objective.

intentional hazard: a source of harm, duress, or difficulty created by a deliberate action or a planned course of action.

IT security goals: the five IT security goals are integrity, availability, confidentiality, accountability, and assurance.

likelihood: an estimate of the potential of an incident's or event's occurrence.

mission consequence: the effect of an incident, event, operation, or occurrence on the ability of an organization or group to meet a strategic objective or perform a function.

model: the approximation, representation, or idealization of selected aspects of the structure, behavior, operation, or other characteristics of a real-world process, concept, or system.

natural hazard: a source of harm or difficulty created by a meteorological, environmental, or geological phenomenon or by a combination of phenomena.

network: a group of components that share information or interact with each other in order to perform a function.

probabilistic risk assessment: a type of quantitative risk assessment that considers possible combinations of occurrences with associated consequences, each with an associated probability or probability distribution.

probability (mathematical): a likelihood that is expressed as a number between 0 and 1, where 0 indicates that the occurrence is impossible and 1 indicates definite knowledge that the occurrence has happened or will happen, where the ratios between numbers reflect and maintain quantitative relationships.

psychological consequence: the effect of an incident, event, or occurrence on the mental or emotional state of individuals or groups, resulting in a change in perception and/or behavior.

qualitative risk assessment methodology: a set of methods, principles, or rules for assessing risks based on nonnumerical categories or levels.

quantitative risk assessment methodology: a set of methods, principles, or rules for assessing risks based on the use of numbers, where the meanings and proportionality of values are maintained inside and outside the context of the assessment.

redundancy: additional or alternative systems, subsystems, assets, or processes that maintain a degree of overall functionality in cases of loss or failure of another system, subsystem, asset, or process.

residual risk: a risk that remains after risk management measures have been implemented.

resilience: the ability to resist, absorb, recover from, or successfully adapt to adversity or a change in conditions.

return on investment (risk): a calculation of the value of risk reduction measures in the context of the cost of developing and implementing those measures.

risk: the potential for an unwanted outcome resulting from an incident, event, or occurrence, as determined by its likelihood and the associated consequences.

risk acceptance: an explicit or implicit decision not to take an action that would affect all or part of a particular risk.

risk analysis: systematic examination of the components and characteristics of risk.

risk assessment: a product or process that collects information and assigns values to risks for the purpose of informing priorities, developing or comparing courses of action, and informing decision making.

risk assessment methodology: a set of methods, principles, or rules used to identify and assess risks and to form priorities, develop courses of action, and inform decision making.

risk assessment tool: an activity, item, or program that contributes to determining and evaluating risks.

risk avoidance: strategies or measures taken that effectively remove exposure to a risk.

risk communication: the exchange of information with the goal of improving risk understanding, affecting risk perception, and/or equipping people or groups to act appropriately in response to an identified risk.

risk control: a deliberate action taken to reduce the potential for harm or maintain it at an acceptable level.

risk identification: the process of finding, recognizing, and describing potential risks.

risk management: the process of identifying, analyzing, assessing, and communicating risk and accepting, avoiding, transferring, or controlling it to an acceptable level at an acceptable cost.

risk management alternatives development: the process of systematically examining risks to develop a range of options and their anticipated effects for decision makers.

risk management cycle: a sequence of steps that are systematically taken and revisited to manage risk.

risk management methodology: a set of methods, principles, or rules used to identify, analyze, assess, and communicate risk, and to mitigate, accept, or control it to an acceptable level at an acceptable cost.

risk management plan: a document that identifies risks and specifies the actions that have been chosen to manage those risks.

risk management strategy: a course of action or actions to be taken in order to manage risks.

risk matrix: a tool for ranking and displaying components of risk in an array.

risk mitigation: the application of a measure or measures to reduce the likelihood of an unwanted occurrence and/or its consequences.

risk mitigation option: a measure, device, policy, or course of action taken with the intent of reducing risk.

risk perception: a subjective judgment about the characteristics and/or severity of risk.

risk profile: a description and/or depiction of risks to an asset, system, network, geographic area, or other entity.

risk reduction: a decrease in risk through risk avoidance, risk control, or risk transfer.

risk score: the numerical result of a semiquantitative risk assessment methodology.

risk tolerance: the degree to which an entity is willing to accept risk.

risk transfer: an action taken to manage risk that shifts some or all of the risk to another entity, asset, system, network, or geographic area.

risk-based decision making: a determination of a course of action predicated primarily on the assessment of risk and the expected impact of that course of action on that risk.

risk-informed decision making: a determination of a course of action predicated on the assessment of risk, the expected impact of that course of action on that risk, as well as other relevant factors.

scenario (risk): a hypothetical situation composed of a hazard, an entity impacted by that hazard, and associated conditions (including consequences, when appropriate).

semiquantitative risk assessment methodology: a set of methods, principles, or rules to assess risk that uses bins, scales, or representative numbers whose values and meanings are not maintained in other contexts.

sensitivity analysis: a process to determine how outputs of a methodology differ in response to variation of the inputs or conditions.

simulation: a model that behaves or operates like a given process, concept, or system when provided a set of controlled inputs.

subject matter expert: an individual with in-depth knowledge in a specific area or field.

system: any combination of facilities, equipment, personnel, procedures, and communications integrated for a specific purpose foundation for a useful transit system.

target: an asset, network, system, or geographic area chosen by an adversary to be impacted by an attack.

threat: a natural or human-made occurrence, individual, entity, or action that has or indicates the potential to harm life, information, operations, the environment, and/or property.

threat analysis: the examination of threat sources against system vulnerabilities to determine the threats for a particular system in a particular operational environment.

threat assessment: the process of identifying or evaluating entities, actions, or occurrences, whether natural or human-made, that have or indicate the potential to harm life, information, operations, and/or property sectors.

threat source: either (1) an intent and method targeted at the intentional exploitation of a vulnerability, or (2) a situation and method that may accidentally trigger a vulnerability.

uncertainty: the degree to which a calculated, estimated, or observed value may deviate from the true value.

vulnerability: a physical feature or operational attribute that renders an entity open to exploitation or susceptible to a given hazard device.

vulnerability assessment: the process of identifying physical features or operational attributes that render an entity, asset, system, network, or geographic area susceptible or exposed to hazards.

Index

Milton Keynes UK
Ingram Content Group UK Ltd.
UKHW040053071024
449327UK00019B/527